HUMAN COMFORT

HUMAN COMFORT

By

KEITH SLATER
M.Sc., Ph.D., C.Text, F.T.I.

Professor of Textile Science
The University of Guelph
Guelph, Ontario, Canada

CHARLES C THOMAS • PUBLISHER
Springfield • Illinois • U.S.A.

Published and Distributed Throughout the World by

CHARLES C THOMAS • PUBLISHER
2600 South First Street
Springfield, Illinois 62717

ISBN 0-398-05128-3

Library of Congress Catalog Card Number: 85-4653

With THOMAS BOOKS *careful attention is given to all details of manufacturing and design. It is the Publisher's desire to present books that are satisfactory as to their physical qualities and artistic possibilities and appropriate for their particular use.* THOMAS BOOKS *will be true to those laws of quality that assure a good name and good will.*

Printed in the United States of America
Q-R-3

Library of Congress Cataloging in Publication Data

Slater, Keith.
Human comfort.

Bibliography: p.
Includes index.
1. Man — Influence of environment. 2. Human comfort. 3. Environmental psychology. 4. Clothing and dress — Psychological aspects. I. Title.
GF50.S57 1985 620.8'2 85-4653
ISBN 0-398-05128-3

For Diane, Hilary and Keirsten.

PREFACE

I BEGIN this book by making the bald statement that human comfort is the most important subject in the world.

You may or may not agree with this sentiment as you start to read the text, but I hope that I may convince you of it by the time you've finished the book. In it, I ask you to think about what comfort means, why it's important, how it affects each of us, how we can measure certain aspects of it, and how it influences our life in the home, at work, in travel, and in everyday living. I also ask you to consider the important part played by textile products in changing our comfort level.

I've tried to write my book in a fairly lighthearted way so that it's readable. If you dislike the idea of levity in a text, I apologise for distressing your high ideals. Put the thing down quickly before you read any further in case I pollute your moral integrity. Then put your money back into your pocket and leave the bookshop at once, thus minimising your own discomfort. What's more, the dirty finger marks will only be on this page, so somebody less intelligent than you will still buy it and I won't suffer the discomfort of remaining unread.

Keith Slater
Department of Consumer Studies
University of Guelph
College of Family and Consumer Studies
Guelph, Ontario, CANADA N1G 2W1

ACKNOWLEDGMENTS

THIS BOOK could not have been written without the help of several people, and I should like to pay tribute to this kindness.

I am particularly grateful to the Master and Fellows of Gonville and Caius College, Cambridge, for electing me as a visiting fellow during the year in which it was mainly written. The abundant comfort provided there, and the library privileges I enjoyed, made the task of working on he manuscript unexpectedly light and transformed the work into a pleasure.

My warmest thanks also go to my friend, Doctor Ruth Tatham, for her patient scrutiny of the manuscript to check the medical aspects, of which I am woefully ignorant. I am indebted to her for suggestions regarding its improvement.

The line diagrams were drawn for me by Mr. Mudd and his staff of the University Engineering Department at Cambridge, and that assistance is gratefully acknowledged. Some of the figures are reproduced by permission of copyright owners, as indicated in the captions, and the kindness in this respect of my colleague, Doctor Anne Wilcock, The Canadian General Standards Board, BMW AG, General Radio, and The Textile Institute, is much appreciated.

Finally, I should like to acknowledge the patience of my wife and family, who have tolerated vast amounts of untidiness and provided large quantities of comfort while I was writing. To all of you, I give my sincere thanks!

CONTENTS

HUMAN COMFORT

CHAPTER 1

THE MEANING OF COMFORT

COMFORT is the most important subject in the entire world. Just stop to think about that statement for a moment. It seems like an outrageous thing to say. Before you dismiss it as ridiculous, though, consider one interesting fact that may never have occurred to you. Every second of your life, from before the moment you were born until the instant you die, you're striving for more comfort; so is every other person on Earth, all four billion of them (or maybe it will be five billion by the time you read this; I write slowly). Twenty-four hours a day, seven days a week, awake or asleep, every individual on our planet, without exception, is constantly trying to increase his or her comfort level. The effort may be either deliberate or completely involuntary. It may be either obvious or totally unnoticed. The comfort increase sought may be of a bodily or a mental kind, but the effort is taking place all the time.

What, then, is this universal need? How do we define comfort? That's not an easy question to answer as you'll realise if you consider it. If I were to ask fifty of my students for a definition of comfort, I'll guarantee that I'd get fifty answers. No. On second thought, I'd get a hundred; students are always in two minds about everything. The point is that comfort is such a nebulous term to define, and is so subjective, that universal agreement on what's meant by the word is almost impossible to achieve. We all know what we mean by comfort but find it very difficult to express our opinions clearly in words that mean the same thing to other people. Paradoxically, we find it relatively easy to define discomfort. If we have a pain, or an itch, or are too warm, or feel afraid, there's no problem in communicating our discomfort, of whatever type, to somebody else, who immediately understands how we're feeling. In other words, it's only when we realise that we're appreciably lacking in comfort, in some form, that we're truly aware of its absence. When we're comfortable, we just don't notice the fact.

How, then, can we begin to approach the subject of comfort? We've got to find a definition on which we can agree, or we'll be talking at cross-purposes. It's obviously impossible for our definition to be quantitative; we can't talk about three pieces (or three litres or three kilograms) of comfort, nor can we think in terms of doubling or halving our personal comfort level. No, any definition we use must be empirical in nature. It has to be based on observation, not on theory, and can't be expressed by the cold reason of mathematical formulae. Let me give you the definition which I first put forward at a conference several years ago; nobody argued with me on that occasion and, as far as I know, nobody has since produced a better one. The definition I propose is as follows: *Comfort is a pleasant state of physiological, psychological and physical harmony between a human being and the environment.* Notice the inclusion of the environment in the definition. As will emerge later in this chapter, the environment has a critical role to play in establishing the comfort level.

Let's inspect my definition a little more closely to identify its strengths and weaknesses. Clearly, comfort has to be pleasant. If there's any unpleasantness of some tangible kind (like a physical pain or an odour), comfort will likely diminish. Similarly, we can't feel comfortable in the presence of intangible unpleasantness existing, for example, when we're quarrelling with anybody or in fear of some unknown lurking danger. Harmony, too, is self-explanatory. The word is defined in dictionaries in terms such as, *the combination of all parts to make an agreeable, consistent or orderly whole,* and that certainly reinforces our mental image of what comfort should include.

Next, we should examine in turn the three types of comfort I've specified: physiological, psychological and physical. The first of these relates to the human body's ability to continue functioning, and very few of us would find it comfortable to exist if that ability suddenly failed. The body has a number of mechanisms which work automatically, without any conscious effort on the tenant's part, in a continuous attempt to maintain or increase the level of comfort. As a simple illustration of their importance, try to do without the most obvious one (breathing) for a few minutes and see if your comfort sensations change at all! The human body is a marvellous machine, which we'll be considering in more detail at a later stage, with many such mechanisms. Their existence is the justification for the apparently extravagant claims made in the opening paragraph of this chapter!

Psychological comfort, the mind's ability to keep itself functioning satisfactorily without deliberate help on our part, is an even more mysterious phenomenon. Scientists are only just beginning to explore the intricacies of the human brain, and we've no real knowledge yet of what's taking place inside it. As usual, it's the absence of comfort which is most noticeable to us. If we're anxious, or afraid, or ill at ease, we quickly recognise the fact. Our mind becomes troubled and, as a result, our comfort level drops. The brain's striving for comfort is automatic, but we can sometimes override the system, just as we can override the breathing mechanism if we choose, for a short time. We can force ourselves to think of unpleasant images or experiences and, if we persist, drive ourselves into a state of illness by so doing. As with the physiological aspects of comfort, we'll be looking more carefully at mental comfort in a subsequent chapter.

Then we come to physical comfort. In many ways, this is the easiest of the three to grasp, since it's more tangible. The sensations of pain when we hurt ourselves, or of choking when a collar is too tight, are too well known to need much discussion; so are the feelings of discomfort arising from an itch or from wearing the wrong clothing, whether the problem is of bad fit or of unsuitable use at a social occasion or because of an inclement climate. Notice, though, that all these examples have in common the fact that they influence one or other of the first two types of comfort. This is generally true as we'll see at a later stage. The aspects of comfort which we term physical almost invariably relate directly to either the physiological or the psychological ones and influence them adversely when we begin to notice problems of physical comfort.

Now, let's get back to this question of the part the environment plays in making human beings comfortable. It's important to realise that you can be perfectly at ease in some situations and totally uncomfortable in others, even though your bodily characteristics wouldn't be measurably different in the two cases. Perhaps some examples will serve to illustrate the point I'm trying to make. Suppose you're swimming in a pleasant pool on a warm summer afternoon. Your body temperature, breathing rate, clothing, state of mind, and any other factor you care to mention all allow you to be comfortable, and you're thoroughly enjoying yourself. Then you dry off, get dressed, and begin to walk home, still in the same frame of mind. The weather becomes cloudy, and rain begins to fall. Soon, you're about as wet as you were in the pool, but you're no longer comfortable. We'll assume that your body temperature hasn't changed and that all the factors which were present in the pool are still in existence here. Yet, in spite of the fact that no changes in any of them can be discerned, you're no longer comfortable. The wet clothes you're wearing aren't supposed to be wet, whereas the ones you wore in the pool were. That's the only difference you can find, but it's enough to ruin your comfort and hence your afternoon.

Then we can consider the situation where you're at a party. You're wearing your favourite pair of blue jeans and a tattered old shirt with paint spills over it, and you find that everybody else read the invitation carefully and wore slinky evening outfits. Even though the clothes are the most comfortable ones you possess, you're still uncomfortable and can't wait to crawl away into anonymity. There's nothing wrong with the clothes, mind you. They fit properly and cover very decorously all the bits that have to be covered. It's just your perception of being out of place that's the problem. I had this very experience only two days ago, which is why it's at the front of my thoughts. I went along to a dinner of a very formal kind and, as a result of my absentmindedness, found I was the only man in the room (of about four hundred, I might add) not wearing a dinner suit. There was even a lady wearing a black bow tie (and she refused to swap it for my exquisite old boys' one!). My best suit suddenly seemed rather shabbier than it had when I put it on. The effect works in reverse, too. The feeling of discomfort would be equally strong if you were the only well-dressed person among a load of disreputably denim-clad guests. It's not what you wear that makes you feel wrong and, hence, uncomfortable; it's the fact that you're the only person there wearing it.

I was once invited at the last minute, as chairman of an international conference at which an important French guest was speaking, to a reception at the French Embassy in Ottawa. My wife, who accompanied me, was wearing an evening dress of fairly plain style, in readiness for the conference banquet immediately after the reception. When we arrived at the embassy, we found to her consternation that the diplomatic circuit had obviously been warned to wear casual clothing that night, and the ladies were dressed uniformly in short (well, not too short; more like knee-length) cocktail gowns. We stayed perhaps an hour, but my wife was relieved when the time came to move on to our dinner. As we left, a latecomer arrived. I won't name her, because she's too well known to too many people, but she gave us a very condescending glance, clearly pitying my wife for having to wear such a plain dress because her husband was too stingy to buy her something better. The lady herself, however, was decked out in the finest satin brocade evening gown that could possibly be envisaged, with a fur stole around her shoulders (it was summer, and the temperature was in the eighties) and diamond tiaras bedecking her head in three or four places (you know how rich those political females are!). The thought of her sweeping into that room full of informally dressed women, the remains of a sneer scarcely fading from her face, kept us in a paroxysm of laughter for the rest of the evening and for many moons thereafter!

Then there are the more obvious examples of how the environment can influence comfort. If you wear the same clothing system throughout the year (unless you happen to be living in a tropical country), you're going to notice different comfort levels eventually. (I'm assuming you wash the garments occasionally; the discomfort to yourself and others will arise fairly quickly otherwise.) Either you'll be far too hot in the summer if you've selected a winter outfit, or too cold in the winter if you've had a summer one for your sole clothing choice. Even within a few days, as I'm sure you're aware, the weather can change enough to make a different type of outfit necessary at many times of the year.

So, the environment is vitally important for physiological, psychological and physical comfort, but we can't easily say how important. To return to this question of the measurement of comfort level, let's consider how possible it is to decide if somebody's comfortable or not and to establish how uncomfortable he is. As I've already said, there's no quantitative method to measure the degree of comfort. Can we, then, define gross limits for the two states — discomfort and comfort — and subsequently decide how much of a change in level is detectable? You're aware from your own experience that you know when you become uncomfortable, but can you tell when somebody else does? Think about it. You're in a room, and the place is steadily getting more crowded. The air becomes hotter by the minute, and you feel as if you might melt. Take a look around you. Most people there are in the same position as you are (there are exceptions because of physiological differences between human beings, as we'll discover later, though we can ignore them for the sake of argument in this example), but can you honestly say that it's obvious? The fact is that everybody, including yourself, continues to behave normally, as though there were no comfort problems, and an observer not aware of the internal conditions of the room would be unable to detect anything wrong by visual inspection.

Eventually, as the temperature climbs still higher, a few noticeable signs do begin to appear. Maybe some faces get red (and not just from the alcohol), or perspiration may begin to form on the forehead or upper lips of the more susceptible guests. In the limit, a fainting fit may affect somebody. In other words, we can disguise our overheated state to some extent by willpower, by acting as though there were no heat, to the point where our body can no longer support our deception. At that stage, the corrective mechanisms of the body begin to operate on a yellow alert, or whatever the physiological equivalent is, and visible signs become evident. Similarly, there are tangible ways of detecting other kinds of discomfort besides that brought about by overheating.

What are these tangible signs? Strangely enough, there are very few ways of telling whether somebody's physiologically or physically uncomfortable but many more to indicate psychological discomfort, the factor of which we know least among the three. Apart from the three I've already mentioned (blushing, perspiraton and fainting), there are no reliable indicators of physiological discomfort, with the notable exception of shivering. All four of these observable effects, however, can have psychological causes, rather than physiological ones; it's not difficult to think of examples of mental stress which could produce any of them. In the same way, the symptoms of physical discomfort, such as pulling in an attempt to slacken tight garments, scratching to relieve an itch, or an involuntary expression of pain, can also have psychological causes. In addition, even the visible indications for psychological discomfort are ambiguous, since it could arise from other sources, so they must be regarded with some caution before you can be absolutely certain of the true cause of the signs you notice. A more detailed look at some of the symptoms will clarify the point.

Let's do it logically, starting at the top and working our way downwards. If people are uncomfortable in a mental, rather than physical, sense, they'll often avoid eye contact with an observer. They appear shifty or continually look away; they frown, have a weak smile on their faces, and generally look insincere or give indications of something incongruous about their facial expression. They may blush, use false laughter, weep, or seem agitated. They may perspire or shiver. An aura of fearfulness may surround them. All these signs, notice, are evident merely by looking at their faces. Now we come to the rest of the body. Their posture may be awkward or they may continually move their hands in a fidgeting manner. Maybe they keep alternately standing and sitting for no obvious reason. They may scratch or rub themselves, bite their nails, and adjust their ties or jackets. Their body movements may give away their agitation as they shuffle around furtively or shift their weight from one foot to the other. If they're sitting, they may drum their fingers on a chair, or tap their toes, or cross and recross their legs, or try to adopt some self-effacing posture. The signs aren't usually all displayed simultaneously, of course; if you see anybody doing everything I've mentioned, you should either send for the nearest strait jacket or sign him on for a television act. There are, nevertheless, endless combinations of expression or gesture which can betray an inner turmoil to an onlooker. Sherlock Holmes would have had a field day as a psychiatrist in modern times!

Even though we can't define comfort by a mathematical equation, then, we can easily recognise it in its absence. Can we identify it in a positive sense? Within certain vague and ill-defined limits, I suppose we can. By concentrating hard on how your mind and body feel, you can assess the state of comfort at which you're presently maintained. By making minor adjustments in posture, or in breathing rate, or by setting the room temperature a little higher or lower, you can learn to detect the slight changes arising in consequence and come to a recognition of your optimal level of comfort. It's possible to make some modifications to this optimal level. When I emigrated to Canada, I found the houses too hot for comfort in winter after my native England, and our guests used to recognise the need for wearing overcoats at the Slater dinner table. Now, I don't find normal Canadian indoor temperatures all that hot, but I just about freeze when I return to the family home in England which used to be so comfortable in the winter.

Before concluding this examination of my definition of comfort, it's only fair to discuss its disadvantages (a ploy which also makes me look like an exceptionally objective scientist). The most obvious of these, naturally, is the fact that it's empirical in nature. That is, it can't be described in quantitative terms and is thus not definable by mathematical equations. This means that we can never compare precisely two different states of comfort. We can say that we're more comfortable in one particular situation than in another, but not that we're twice as comfortable.

Secondly, it begs the question of what is meant by "pleasant" and "harmony," terms almost as difficult to define as "comfort." Finally, since it includes a human being, with all the variability which that implies, we can expect a lot of disagreement about the comfort level of any particular environment. There is also the point that we should distinguish between the external and internal environments of the body by using environment in a total sense, a distinction which gives us an unusual way of examining the effects of environment (both mental and physical) on comfort. Despite all these drawbacks, however, the definition I provided does serve to give us a starting point from which we can get some kind of grasp on the subject and on the way it affects us.

The human body, then, is an adaptive one, which can get used to a range of comfort levels and which can disguise an unpleasant reduction in comfort from a casual observer to a considerable extent. When the limit of its capacity to deceive is reached, unavoidable signs become evident and are beyond the control of the individual displaying them. Comfort, which we take so lightly for granted, has changed to discomfort and, in the process, forced the new situation insistently into our consciousness!

CHAPTER 2

THE IMPORTANCE OF COMFORT

NOW THAT we've got some idea of what we mean by comfort, or accept that we may never know precisely what it is, we can begin to discuss its importance in everyday life.

The first irrefutable fact to note is that comfort does indeed affect all aspects of our existence. Even before birth, the human embryo is attempting to increase its level of comfort. Ask any mother what the unborn child within her body did to her for the few months before she gave birth to it. She'll tell you of the internal kicks she received or of the pains which would suddenly disappear as the baby found a better way to lie. All that embryonic motion is a result of the foetus's attempt to find a more comfortable location for itself. Almost at the instant of birth, the baby gives its first cry to express displeasure at being wrenched away from the comfort of the womb. Psychologists are only just beginning to realise how traumatic a step that must be and are recommending, for instance, that the mother should keep the baby close to her body so that her familiar, comforting heartbeat is still audible to reassure the newborn infant that the separation isn't an end to the security known hitherto.

When a baby grows, its comfort demands become even more intense. If an infant is dirty, or hungry, or sleepy, or lonely, try to deny it the fulfilment of its needs and see what happens. There's nothing quite so pathetic as the cry of an unhappy baby, as any parent dragged out of bed to feed and change a howling offspring at two in the morning will readily acknowledge. It certainly changes your own comfort level! You just can't resist the noise, and the baby seems to sense your response from a very early age. The human infant is possibly the most selfish, manipulative, unreasonable creature in the entire realm of nature!

A child, too, will try to maximise its comfort. Ask your two-year-old to go to bed without his or her favourite toy and then be prepared for a long night! When your offspring finally does go to sleep, see how the toy (which you finally found) is clutched tightly. Notice the thumb in the mouth or the hand fiercely clasped round the piece of blanket that feels so soft.

As the child gets older, the needs change, but they're still there in some form or another. Friends, the familiarity of established routines, and favourite clothes or meals all serve to reinforce the feeling of security on which the healthy growth of the child depends. The teenager, though probably never willing to admit the fact, doesn't escape the need for comfort. The wrong brand of jeans for current fashion demands can drive my own teenagers into a frenzy of anguish if they have to be resurrected for emergency use because the current model is dirty or in want of repair. Makeup that doesn't quite match the colour of a boyfriend's dog's teeth, or something equally unfathomable, can ruin the wearer's entire future life for almost ten minutes. When my own daughters were somewhat younger, any one of them would literally cringe with dread if she happened to see me on the street, in case I should speak to her in front of her friends. The friends in question, naturally, were very warm in greeting me and took for granted my own offspring's need to skulk behind them until I'd passed. The discomfort of being ashamed of one's relatives' appearance, face, clothing, etc., takes a long time to overcome when you're young!

Even the toughest of adults need comfort. Take the most macho man of all the men you know and ask him to relinquish his cigarettes or his beard or any of the other signs of manhood which he's so carefully cultivated. You might as well ask him to put his finger in his mouth, stand in a corner and cry, because there's no discernible difference in his horror at either possibility. As we get older, the needs increase rather than fade away. Think of your own feelings if the car won't start, or if your coffee's gone cold, or your favourite chair's been taken by that old bag who comes in to talk to the wife and here you've been slaving away at the office all day and you just want to come home for a bit of peace and . . . need I go on?

What about the nylons that have a run? Try persuading any lady of your acquaintance that it doesn't matter if she wears them to the party. When she gets there, naturally, her brand new dress looks exactly like the one that younger girl bought on sale at half the price the day after she herself did. You can't win. Accept the fact that you need comfort throughout life and seek it wherever you can find it.

In old age, the demands intensify. The favourite pipe has to be available. The newspaper must be delivered on time. The meal should have been cooked the way you always have it done. The children and grandchildren ought to come on the right day and stay for the proper length of time. People should stop fussing about you and should show a bit more interest instead of ignoring you, both at the same time. Finally, as life draws to a close, the peaceful quiet of old age should be allowed to settle on you. There should be no disturbance of your customary routine. Solace of a spiritual and physical kind must be there when you need it but not thrust at you unasked. At the very end, the comfort of a friendly hand, or an attentive nurse, or a welcome painkiller, is all that you crave. From cradle to grave, for old and for young, for rich and poor alike, comfort is the major need in life and to the edge of death.

Notice that the comfort can be mental or physical. Both are important, and it's impossible to decide in advance where the greatest need will lie. We need a balance of all three categories of comfort, and we need them in direct response to the events of a specific situation. Wherever the discomfort exists and from whatever cause, there is where the remedy is most urgently sought. Without comfort, human existence loses its quality, and we say that life isn't worth living if we're permanently in an uncomfortable state.

On occasion it may quite literally be true that life can't be prolonged without comfort. Comfort can be stretched to the point where it involves safety, and, in the limit, the absence of comfort can mean the difference between survival and extinction. Consider the most obvious instance of discomfort which commonly affects many people, the situation where you feel cold. If there's a chill in the air, you may feel uncomfortable and be forced to walk more quickly (and hence get out of breath, another manifestation of discomfort) to remain warm enough to restore some measure of thermal comfort.

Now translate that to the case where you're walking home on a bitterly cold evening, faced with a blizzard, and forced to go on foot because the weather's too bad for traffic to proceed. As you struggle to make headway against the fierce wind, fighting for breath, you suddenly realise that you've lost all sensation in your ears. There's a risk of frostbite, you recognise, unless you can get to shelter. Suppose, though, that there's no shelter before you reach home, an hour or so away in this weather. You're in danger of needing medical attention to restore your ears to health and may be forced to suffer pain for some time before the trouble's cured.

Now imagine the setting of your blizzard switched to the Arctic and the time between you and safety increased to several hours. The situation has become critical, even life threatening. If the frostbite is severe (and especially if it affects the more remote parts of your body), you may need an amputation because gangrene has set in. In the limit, your frail strength may not be enough to carry you to the refuge, and you may just collapse to lie down in the snow, never to rise again. What was originally mere discomfort has become death. The environment is indeed important as we noted in the previous chapter.

You can think of many other instances where comfort becomes tantamount to suvival. The infected sore or blistered ankle may be a source of minor irritation at home, but it becomes much more critical if you're climbing a Himalayan mountain and are out of reach of medical supplies. Swimming at the beach on a chilly day can be bracing, but the same water temperature when you fall overboard on a dark night from a ship in the North Atlantic can give you a different outlook on life (or on what bit there is left of it). A dry throat as you sit by the poolside bar is easy to rectify, but there aren't too many pools in the Sahara if the same discomfort happens to strike you there. Running in a race may make you breathlessly uncomfortable, but if you're fleeing from a lynching mob your shortage of breath could very easily become more permanent. You may be unhappy about the solitude if you're on your own during a long walk, but it could be worse in solitary confinement in a South American prison. The noise on your teenager's radio may be laying all kinds of stress on you, but an explosion just outside your left eardrum might conceivably be even less pleasant. The bright light shining in your eyes makes you feel uncomfortable, but a laser beam would burn the retina away and leave you blind. You may find the taste of the food you eat unpleasant, but the reaction to salmonella or to an allergenic dish could be far worse. The smell of the local sewage plant makes you wrinkle up your nose, but a whiff of prussic acid gas can wrinkle up all the other bits of you as well. You may be annoyed about the vibration you're experiencing in your car, but if you crash it at high speed the bump can shake you into little pieces. I could go on, but I'm sure you get the gist. Comfort, in the limit, means survival.

We've been talking about bodily needs in this little excursion to the limits of comfort. What about mental comfort? Is there an equivalent to the risk of physical damage or annihilation? The nearest one, I think, lies in the mind's ability to protect itself against insanity, the state in which the brain has ceased to recognise reality as defined by most of the worlds' other inhabitants. Again, the discomfort of a mild disorder becomes, as its level increases, a threat to the ability to function well or, in the limit, at all. Fear of the dark, or of the unknown, is a normal reaction at times, particularly if you're very young. But if the fear begins to encompass more and more of your existence, then phobias, or even paranoia, can result. When every single aspect of thought has separated itself from reality as others know it, then the integrity of the mind itself is threatened and the situation has become grave indeed.

Fortunately, few of us are ever thrown into that pit of horror, but that doesn't mean we don't experience mental discomfort. We need to be reassured that our actions, or our behavior, or our appearance, are acceptable to ourselves and to others. We need the self-satisfaction of feeling right in our clothes or of knowing that they're the right colour and style. We have to be assured that all the colours we're wearing match properly; that the entire outfit is coordinated as a unified whole. We shun the disfigurement of a scar, or a cast in one eye, or any bodily deformity, and have it corrected surgically if possible. Even something as small as a pimple, or an eczema rash, can upset us out of all proportion to its size or importance to health. A speech impediment, such as a stutter or an inability to pronounce certain consonants properly, can be devastatingly embarrassing, again particularly for a young person. We've already dwelt on the discomfort experienced as a result of wearing an unsuitable clothing system and will return to it in more detail (as we will to all these aspects of psychological comfort) in due course.

For the present, though, I think I've given you some idea of how important comfort can be to humanity. It's vital to us in or before birth, thoughout childhood, and during our entire lives, right to the very end. It enables the ego to maintain a healthy self-image within our being. The absence of comfort can influence safety and can even affect the chances of our surviving if it's witheld in critical situations. Physically and mentally, in differing degrees, comfort matters to each of us all the time.

CHAPTER 3

PSYCHOLOGICAL ASPECTS OF COMFORT

PERHAPS the most difficult type of comfort to appreciate is the mental one, mainly because anything concerning the mind is concealed from the view of others. We don't know what's taking place in somebody else's head (and not always in our own!), so we've no sure means of assessing any thoughts except those which we ourselves are generating.

For that reason, it might be a good idea to tackle this aspect of comfort, the psychological one, first in order to dispose of it. In the limit, after all, it's in the brain that comfort or discomfort is ultimately registered, so it's only natural to consider what's happening in the brain at the outset. We can't expect to be comfortable in our clothes if we're not comfortable in the place where comfort is sensed. It's like taking care of a baby, when you think about it. There's not much point in putting a clean dress on the baby if it needs changing; the smell won't go away unless you do the thing properly.

So let's look in some detail at the psychological aspects of comfort. What are the factors which must be present for a human being to have peace of mind? The first factor of importance, surely, has to be the self-image possessed by each of us. If we can't live with ourselves, we can't possibly hope to be comfortable with others, and if mental comfort is absent no amount of physical comfort will be able to reconcile us to our existence.

Once that's accepted, the relationship between an individual and other people must be considered. The human animal is a gregarious one, needing the companionship, or at least the company, of other creatures of the same species in order to thrive. We need to impart our ideas, our feelings, or our problems to people who are prepared to listen. In return, we have to be prepared to listen to their discourse, for they, too, have needs. Yet, there are times when we must be alone to allow ourselves the luxury of private thoughts. We need to be free from the interruption of other voices intruding upon our mental processes. Clearly, some kind of balance has to be struck between these two conflicting situations.

We can identify three types of privacy. The first of these, and the most obvious, is solitude. There are times when a human being needs to be utterly alone in the world. Technically, that's a little tricky, because the world isn't the kind of place you can reasonably expect everybody else to leave for a few hours. The difficulty can be resolved by creating an appearance of total solitude. We go for a walk in the country, or climb a mountain, or set sail on a dinghy. We can even achieve solitude by enclosing ourselves within metal and moving around in the self-imposed prison we call our automobile. As long as we can deceive ourselves into thinking we're totally alone in the space surrounding us, we're deluded into the belief that our solitude is complete. It's only when we're rudely awakened from this delusion that our comfort disappears.

Consider the thoughts passing though your mind when you find that your favourite hiding place has been penetrated by someone else. You reach the top of the mountain, only to find a picnic group with plastic bags strewn round them and realise, to your chagrin, that they've driven up via a back road you didn't even know existed. Or what about the time when you're coasting along happily in the car, lost in your thoughts, and some idiotic driver cuts in front of you, forcing you to brake or swerve? Doesn't your comfort level diminish suddenly? Not just because of the emergency aspects of the situation, either. The same reaction exists when you're forced to wait behind a driver who is slow to get away after the light has turned green. Psychologists often puzzle about the change in a human being's nature when he gets behind the wheel of a car. I'd like to suggest that at least part of the aggression displayed arises as a result of the need to safeguard mental comfort and the fear that this comfort will be violated.

The second type of privacy is found in silence. We can be in a room full of people and, as long as they're not speaking, we can be persuaded that our need for privacy is met. Even if some of them are involved in conversation with others, our privacy isn't violated as long as we're excluded from the conversation. Think of the situation in a doctor's office. Several people sit around, waiting their turn to consult with the doctor. Most, if not all, of them are there because there's a health problem. That alone is sufficient to give rise to a certain amount of disquiet and, hence, discomfort. They don't want any more laid on them, so they refrain from talking. Each person sits there, wrapped in thoughts of his personal troubles, needing the privacy to think about the treatment ahead. Conversation is stilted at best and, more usually, totally lacking. In a library, where problems of a different kind are being resolved, the same kind of situation prevails. The occupants are wrestling with a difficult essay or seeking information which isn't too easy to understand for them, and the last thing they want is to be interrupted. Silence is golden indeed when privacy is at stake!

Finally, we come to the third type of privacy, that of anonymity. We can be in a crowded street, surrounded by people who are talking loudly into our ears, but if they don't know us we feel that our private thoughts aren't violated. Even at a party, where we're accosted by total strangers, we're insulated from the mental contact which threatens our psychological security as long as we can remain strangers. It's only when the contact develops beyond this stage, to the point where we're exchanging thoughts dredged from below the conscious surface, that we begin to be on our guard against saying the wrong thing. As long as the conversation revolves around the weather, or the latest fashions, or the world situation, we remain comfortable. Only when the discussion turns to more personal matters, in which the risk of some relationship developing becomes a real one, do we begin to choose our words more carefully. The situation has changed. We recognise that our mental comfort is reduced, or likely to be reduced, by the direction the conversation is taking, and we begin to take steps to safeguard ourselves against any potential discomfort which might arise.

How do we search for privacy? The ways I've already mentioned are effective, but what happens if they're not practical? In the middle of a large city, with a busy work schedule, it's not always possible to find a handy mountain to climb. There may not be time to go to the library for a couple of hours. Not too many large cities have lakes big enough for a sailboat running down the main street. In days long gone, there was always the option of leaving the everyday world to become a hermit or a monk, but those possibilities are now the exception rather than the rule. The chance of becoming a recluse in the Howard Hughes class is reserved for the few who belong to the Howard Hughes millionaire club. It costs money to live in these hard times. Society doesn't willingly trek up to the mouth of the local cave with a bowl of gruel. The modern equivalent of the poor hermit (the hippy or the welfare bum) isn't a respected member of society. The religious orders, too, are reluctant to accept all and sundry. It's not quite as easy to get into the monasteries as it used to be. You've got to be devout first, and most people aren't willing to make that commitment as a means of escaping from the cares of everyday life.

So what are the modern ways of coping with the need for privacy in a busy world? The use of the car as a means of insulation from life has already been mentioned. When we're encapsulated within our vehicle, we're taken out of the world into a private place of our own. Nothing can touch us. Nothing can disturb our peace. Comfort is ours for the duration of the ride. Our subconscious mind, recognising the need for privacy, takes over the mechanical task of operating the vehicle. How many times have you driven along a road for several hours, then arrived at your destination without any memory of the journey? A car is a mechanical device which has certain mechanical needs to be fulfilled, and the reflex area of your brain supplying these needs can do so without any conscious volition on your part.

Then there's the human being who chooses to be "loner." He sits on the fringe of the group at a party or goes off for a long walk by himself frequently. Maybe he retires to a quiet room in the house, away from the rest of the occupants, or buries himself in a newspaper at the breakfast table to cut himself off from the risk of having to speak to the people sitting with him. He refuses to be drawn into conversations, preferring to sit listening rather than take part. The isolation of his own company is preferable to the small social chitchat of a group with nothing of great moment to say. I regard myself as something of an expert concerning such people, mainly because I'm one of the world's prime examples of them when I choose to be. Most of us have a part of this type of behaviour in our makeup, but the proportion varies enormously according to the need for privacy. As I continue to tell my wife when she remonstrates with me for my anti-social attitudes, psychological comfort is necessary to maintain our sanity!

The need for privacy of a somewhat different kind is manifested in the concept of personal space. Each of us is surrounded by an invisible, intangible barrier which defines the limit of approach of another human being. The extent of this barrier is indeterminate, in that it can change with circumstances, but it's almost always present. If we're in a bus with many other people, there's no mental discomfort until overcrowding reaches the stage at which those around us violate our personal space. We're prepared to allow the boundaries to be diminished to some extent for the sake of expediency, but there comes a time when we feel theatened by the close proximity of the person next to us. Watch somebody struggle to reach the door of a bus in which people are packed tightly together. His progress through the ranks of bodies is marked by tiny gaps which appear like magic as each person in turn shrinks away from too intimate a contact with the departing passenger. The sense of relief as you get off a bus as full as that tells you, only too clearly, that you were more uncomfortable than you were prepared to admit while you were aboard it.

The variation of the extent of an individual's personal space is governed by many factors. For a situation where crowding is necessary (as on the bus), our minds are prepared to accept a compromise. We're not comfortable, but we recognise the need for self-sacrifice to some extent in the common interest. After all, if everybody insists on the correct personal space, we may be the ones who are left behind at the bus stop! The person next to us can also influence the interpersonal distance we need. If it's a complete stranger, we want to keep the gap between us wider than it would be for a good friend. The more we trust or love somebody, the closer we're prepared to allow him or her to approach us. This is a very convenient arrangement; without it, world population would tend to decline in a most alarming manner.

Mention of which brings me to the point that racial or ethnic factors may operate on the establishment of the correct interpersonal distance. As far as we can tell, the oriental races prefer to be in much closer proximity than do the occidental ones. There's some disagreement among psychologists about the cause of this difference. Some experts maintain that it's the result of hereditary factors, while others ascribe a purely environmental explanation to the phenomenon. The cause is immaterial to this book; it doesn't much matter whether the barrier is based on genetic factors or whether it arises from the degree of overcrowding of the society in which we're raised. What does concern us is the effect which the distinctions have. There's nothing quite so funny or so embarrassing, depending on whether you're a spectator or a participant, as the sight of two people with appreciably different interpersonal space needs pursuing each other round and round in a circle as they converse. The person who likes to be close to his companion moves in and, after a few moments, the other person moves away. The first one moves in again, and the second one retreats, then the whole sequence is repeated endlessly throughout the entire period of time in which they're together. After they separate, one of them has the distinct impression that he couldn't get near his companion, and the other feels that he was constantly being crowded.

The differences aren't always ethnic; two people of the same race can have different space needs and take part in the same dance routine. You've only to travel in an elevator to be made aware of this fact. A group of people, all talking freely to a companion, can converge at the door of an elevator and you'd think the national anthem was about to be played. As they pass into the enclosed space and see how crowded it will be, a hush descends on them. They push their way in, move as far back as possible, face the doors, and stand gazing in rapt attention at the little numbers that tell you when you'll be released from the confinement of this trap. There's a sensation of shrinking away from those around, as each person tries in vain to avoid trespassing on his neighbour's space.

Then the situation changes. In comes somebody who's unaware of the correct way to behave in an elevator. He continues his conversation, complete with gestures if necessary, to the obvious discomfiture of his companion, who has been properly educated in these matters. The tension builds. All eyes are averted even more. No one wants to admit that the intruder is even present, let alone talking, and the atmosphere remains awkward until the doors open to discharge him, still talking, to the obvious relief of everybody in the vicinity.

Let's now turn to other types of mental anguish. The first one to spring to mind is the converse of the need for privacy, that is, the desire for companionship. Loneliness can be a devastating experience. If we're alone in a large city, surrounded by hundreds of people who seem determined at all costs to maintain their privacy, we experience a psychological discomfort more miserable than almost any other. As I mentioned earlier, man is a gregarious animal. He needs the presence of other members of his species for communication. There's nothing quite so frustrating as to be surrounded by people who won't, or can't, talk to you.

Even though I'm one of the world's loners, I've often experienced the feeling of isolation which comes from having nobody who'll speak to me. My work as a textile scientist takes me to various parts of the world on conferences or other business engagements, and I've often felt the need to talk to somebody but had that need frustrated. Partly in an attempt to overcome this problem, I've managed to teach myself to speak three foreign languages. And, though that alleviates the difficulty to some extent, it doesn't completely cure it. I can enter into casual conversations with strangers but only to ask for directions or to order a meal. Being able to communicate verbally doesn't necessarily mean you can communicate emotionally, because communication is a two-way process and needs somebody at the other end of it from you.

You're aware of this fact, I'm sure, in your own personal experience. If you're away from home or friends for a period of time, even in a place which you know and where you understand or speak the language perfectly, there is still the same feeling of desolation. It increases as the distraction of other activities fades. It's easy to thrust it out of your mind when there is a task to be accomplished, but it's there in greater strength when the day draws to a close and there are fewer sidetracks to follow. You've only to go into the bar of any hotel in any fair-sized town to find ample proof of the way in which loneliness can strike. On any given evening, it will be full of tired-looking businessmen sitting in solitary misery with a drink to help deaden the pain. Every one of them is there to seek the companionship of anybody who'll talk to him. If she happens to be blonde and amenable to the blandishments of flattery, so much the better. The wife never did understand his needs the way this girl seems to!

Then there is the mental discomfort of routine problems that affect everybody at some time or other. Financial worries may be inflicted on the long-suffering mind. If there are concerns about where the next meal or the next month's rent (or even the next million dollars) will come from, comfort is lost. If there are family problems of an emotional kind, the anxiety is there to gnaw at us. The spouse, or the child, who's having an upheaval of some kind arouses within us an emotional reaction in sympathy and gives rise to a consequent degree of discomfort.

Illness or bereavement, either on our own part or for somebody else who means a lot to us, can be a prime source of anxiety and, hence, of discomfort. A sickness which afflicts us always seems worse when we don't know exactly what it is. How many times have you heard people say they don't care how bad the illness is as long as they know the truth? They say it's not the illness itself that really bothers them but the fact that they can't be certain whether it will or will not eventually be cured.

Even a physical problem which is troubling somebody can bring on psychological discomfort. Consider the last time you suffered a pain which, though clearly only temporary, was severe enough to be felt continually. Perhaps it was a toothache, or a headache, or maybe you suffered a sprain or were badly bruised in a fall. The mental anguish of being unable to concentrate, or of being restricted in some activity, is often worse than the actual physical pain.

I love to eat, and there's nothing makes me feel more unhappy, on the mercifully rare occasions when I suffer from toothache, as the fact that I'm prevented from eating a hearty meal by the pain in my mouth. In the same way, I was in a fine state of frustration when I was unable to move around freely on one occasion as a result of an accident I suffered in a football game. The torn ligaments in my knee gave me no pain as long as I used the crutches properly, but the sheer resentment of being unable to walk normally, dress myself quickly, or carry out any of the similar activities we take so much for granted, nearly drove me mad in the months the injury took to heal.

When the pain becomes worse, mental strain correspondingly increases. I count myself as fortunate indeed that I've not yet been afflicted with any chronic illness. I can visualise only too well the misery of having to cope with pain on a daily basis. So many people are routinely forced to accept the agony of an ulcer, or the misery of a degenerative disease. Others must live with the knowledge that they're suffering from a terminal illness and have only a limited amount of time left before it will inevitably conquer them. Even if there's no physical pain, the mental one, the anguish of the unknown, is constantly there and is often worse in its effect. The feeling of helplessness, of inevitability, must be a tremendous burden to carry.

Fortunately, the mind appears to be able to heal itself to a certain extent. The records of medical science tell us, again and again, of cases where human beings have conquered the fear of death or have been able to learn to cope with a disability which ought, by all reasonable rights, to have destroyed them mentally. Somewhere, deep inside the brain, is the ability to heal mental wounds, just as the body has the ability to heal physical ones. A human being is able to strive for an increased comfort in the psychological sense as well as in other ways. This struggle takes place without any help on our conscious part, though we may alleviate it to some extent by appealing for external help from either religion or philosophy. We shall be looking a little more carefully at these two aids to comfort at a later stage, but it's obvious from the experience of our own or other people's lives that they can be invaluable aids to the process of restoring some degree of comfort to troubled minds.

Finally, there are the types of psychological factors which cause us no lasting physical discomfort but which can haunt our memories for years. The embarrassment produced in us by awkwardness, or by our appearance, can be devastating at the time and can be a source of mental discomfort on occasion for the rest of our lives. I can still remember blunders I made as a teenager, or as a student, which bring the most peculiar sensations of anguish to mind whenever I recall them. If I did the same things now, I'd pass them off lightly without a thought, but at that time it almost seemed as if the end of the world should logically be announced by gaffes of such magnitude!

A physical peculiarity, unnoticed by anybody else, can give rise to emotional upheavals out of all proportion to its importance in the mind of the "afflicted" individual. You must have known friends who suffered agony about a nose that was too big or a mouth that was too small, even though it looked perfectly all right to you. We all have our own pet dislikes about our own appearance, and I'll make a pact with you: I won't tell you mine if you don't bore me with yours!

The mental chagrin of social incorrectness can also bring about misery to the victim. If we're unable to fit in with our surroundings, then we're mentally uncomfortable. It may be a case of wearing the wrong clothing, as with my wife at the embassy reception, or it could be a matter of cultural difference. If you're eating at the home of an acquaintance from another culture, you're always a little on your guard against doing something unforgivably offensive. The difference between British and North American manners has occasionally caused us some discomfort on a very small scale, and it's easy to see how that could be magnified in a situation where greater differences in acceptable behaviour exist. Eating, speaking, and even such harmless (to our way of thinking) pastimes as looking at somebody can be viewed very differently in other parts of the world. If you don't believe me, try asking a lady to remove her veil in the areas where purdah is still practised!

A handicap which disables in some way, either physically or mentally, is another example of a factor which can give rise to psychological discomfort. The inability to behave in a manner akin to those around you is distressing, and discomfort exists wherever there is stress. This type of handicap, like those of social conformation, is connected with the need for anonymity, the basic need for privacy, which brings us back full circle to where we started.

One very important fact needs to be brought out before we conclude this chapter, though, and it's this: although there may be discomfort on the part of the person suffering from the problem, whether social or physical, there's often considerably more discomfort on the part of others present. Again, we come back to the point about the importance of the environment. The surroundings, or the other people who are around, may seem to be completely irrelevant, but they're often important. The discomfort may affect only the immediate victim of the problem, may be present also in the minds of the beholders, or may be of concern to the bystanders only.

You must have been in the situation of standing helplessly watching while somebody with a physical handicap struggles to accomplish some task which is so easy for you but virtually impossible for him. Nevertheless, you recognise that he must accomplish it himself and that your uninvited help would be received with animosity, let alone refusal. The man struggling to carry out the task goes about it with gusto, seemingly oblivious of the fact that he's the centre of attention. He's used to his handicap and has learned to cope with it. His brain has already healed the mental discomfort which originally afflicted him. Now you're the one suffering embarrassment, as you watch him strive to achieve his aim, while he is perfectly comfortable in his mind. He may be at a considerable handicap in the opinion of most of the bystanders, but it's they who are suffering most, in terms of psychological discomfort, as they stand there watching him!

Mention of the act of watching something brings me to the point that discomfort can have a purely visual origin. If we're forced to look at something unattractive, it gives us an uncomfortable sensation. The sight may be one inspiring horror, as on many of today's entertainment screens, or the result of a car smash. Alternatively, it may be something as simple as the face of somebody we dislike or a ruined view. One of my pet hates in this category is the vista along the Champs-Elysées in Paris, looking from the Louvre towards the Arc de Triomphe. Some architect with a criminal callousness has built, several miles away, two apartment blocks which are framed exactly in the centre of the magnificent arch, ruining the view along what was once the most attractive urban avenue in the world! Truly, the visual comfort is easily destroyed by simple thoughtlessness.

Thus, we come to realise that the topic of comfort is an incredibly complex one. Psychological comfort, in particular, has so many layers of meaning that it's impossible, in a general book of this breadth, to deal with them all. In many cases, there is no logical reason why such discomfort should be present at all, but the human being, whatever he might think, isn't a logical creature. If you want to concern yourself with logical activities, close this book now, forget about anything to do with human beings, and go to play with the computer in your den!

CHAPTER 4

PHYSIOLOGICAL ASPECTS OF COMFORT

ASPECTS OF comfort which are physiological in character can be identified in all parts of the body and in virtually all of its mechanisms. Almost exclusively, the body's actions associated with this area are involuntary in nature. The cardiovascular system, the skeletomuscular system, the central nervous system, the pulmonary system, the digestive system, the sensory faculties and the thermoregulatory mechanism are all involved in some way with the body's continual attempt to become more comfortable. Small wonder, then, that I dared to make the claim, at the beginning of this book, about the importance of comfort! Let's examine each of these areas of the body's operating controls in turn to see how their function relates to human comfort.

I suppose the first thing most of us think about when we consider how a body works is the cardiovascular system. That is usually the bit that ultimately gives up when it's time for us to cash in our chips to keep the appointment with the Grim Reaper. No matter what illness carries us off, we normally come to the last straw when our heart ceases to beat. In addition, heart attacks are notorious as one of the major causes of fatality for humans, at least in those parts of the world which we laughingly refer to as civilised. It should come as no surprise, then, to learn that the cardiovascular system plays an important role in determining comfort level for the poor mortals inhabiting this planet. Right at the outset, of course, we must accept the fact that all of the body mechanisms are intimately related to each other and, therefore, instrumental in enhancing or reducing the comfort of other mechanisms. In addition to that simple relationship, though, the cardiovascular system itself has a direct bearing on comfort.

This system, which is illustrated schematically in Figure 1, operates automatically and consists essentially of the heart, the arteries and the veins. Blood containing impurities absorbed from body cells is pumped from the right ventricle of the heart through

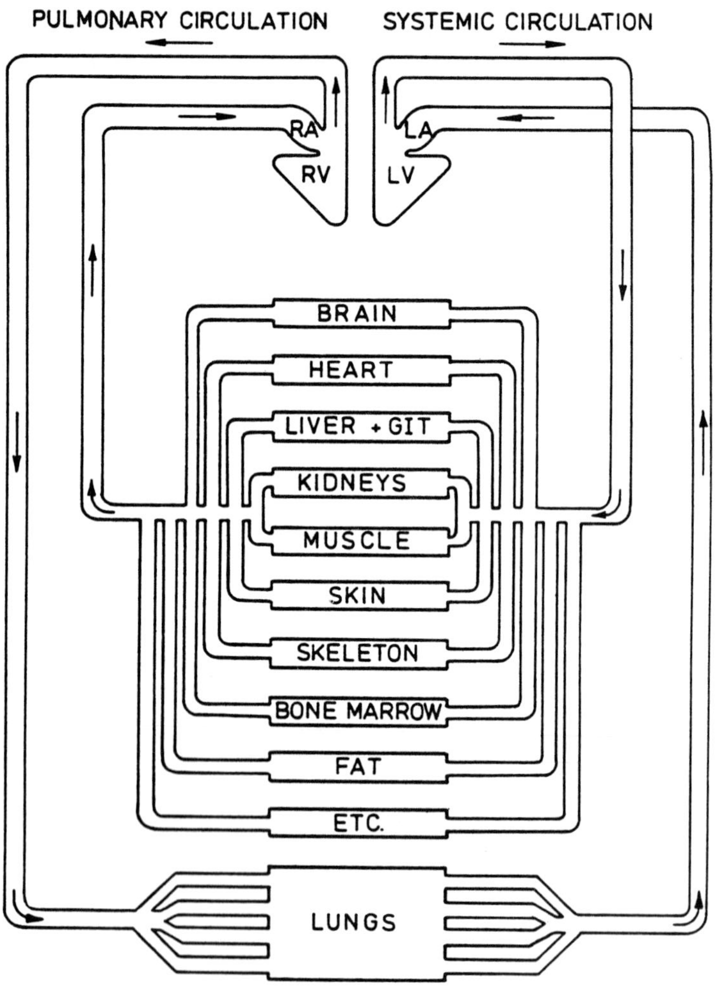

Figure 1. Schematic diagram of the human cardiovascular system.

the lungs, where it absorbs oxygen and gets rid of carbon dioxide, together with the other impurities. The purified blood is then returned to the left atrium of the heart and passes through a valve into the left ventricle. From there, it's pumped around the body, via the arteries, delivering to each cell the oxygen it needs to continue to function. At the same time, carbon dioxide, together with other waste products, passes from the cell into the blood. This "impure" (oxygen-depleted) blood then returns to the right atrium of the heart via the veins, and passes through another valve to the right ventricle, to be sent from there to the lungs for the next cycle of operation.

In the body, there are a large number of circuits in parallel. In each of them, the arteries damp out big pressure fluctuations and the arterioles (which can be regarded as miniature arteries and will be mentioned again in due course) act as taps to regulate the flow rate as needed. Capillaries and thin venules (miniature veins, also to be mentioned later) act as exchange sites, allowing material to diffuse across their walls in both directions. The veins are capacity vessels, acting as adjustable reservoirs to meet variations in volume and distribution needs. It's interesting to note that the four activities of damping, regulation, exchange and storage are shared to some extent by all the blood vessels, but that a kind of specialisation has developed over evolutionary time, so that one activity tends to predominate, as described, in each particular type of blood vessel.

The most important function of the blood flow, however, is the nourishment of the brain. Most of the other parts of the body can survive for about half an hour (or longer in some cases) without adequate oxygen, but the brain suffers irreparable damage if deprived of this vital element for four to eight minutes, depending on such factors as temperature or cell age. There's also the matter of the survival of the body as a whole. As we'll discuss at a later point in this chapter, the human organism is perfectly capable of sacrificing bits of itself to ensure that the body can survive if oxygen depletion occurs. If necessary, odd parts are killed off (and must subsequently be amputated) in order to make sure that all possible reserves are retained to keep the human being from extinction. Fingers, arms, toes, legs, and so on can be dispensed with in dire emergency, as long as the body's vital organs aren't destroyed, but this is emphatically not the case where the brain, the stomach, the kidneys, the lungs, the liver, and a few other bits of offal are concerned. We shall shortly be dealing in some detail with the question of how the body guards itself against large temperature changes, the usual cause of a need for emergency self-protection, but there is another aspect of comfort more directly connected exclusively with the cardiovascular system.

This relates to the question of how the system ensures that it can continue to function. There are two factors to this problem. The first of these is the quantity of blood available (and the quality of that blood), and the second is the ability of the precious fluid to reach the places where it's needed. The quantity can be decreased, for instance, if large amounts of blood are lost in a wound, by injury, as a result of burns, or by major surgery. Then the body goes into a state of shock, forcing the victim to stop any appreciable activity, until recovery procedures can begin to go into effect. Apathy, weakness, shallow breathing, rapid heartbeat, feeble pulse, lowered blood pressure and cold or clammy skin result and, in severe cases, fainting occurs. In the limit, death can come about, but the body is trying to stave off fatality for as long as possible by mustering all its reserves to deal with the crisis. That is why no other activity is tolerated by the automatic system of protection and why the body in shock disables itself from any strenuous behaviour.

The quality of the blood is maintained by the vital organs charged with this task, mainly the liver and the lungs. The liver is the largest internal organ, weighing up to 1500 g (or 2% of total body weight) in a human adult. In addition to its function in helping to digest food by synthesising essential proteins and breaking down carbohydrates, fats, alcohol, or various other compounds carelessly ingested, it acts as a storage site for glycogen, vitamins and iron. It's also the site where many foreign substances, such as drugs or bacteria killed by the white blood cells, are removed from the blood, thus maintaining its ability to support the body cells. As with shock, the body often has a means of attracting attention to a damaged liver (through severe pain in this case), though much liver damage is, unfortunately, silent.

The lungs facilitate the necessary free exchange between oxygen and carbon dioxide, which permits us to survive, and will be discussed in more detail in the next paragraph. The blood's ability to reach places is governed by the condition of the heart, working as a pump, and of the arteries, acting as the supply tubes. If either of these two parts of the body fails, pain is again present to give warning of potential disaster. The sharp squeezing sensation characteristic of angina is often a preliminary warning of impending trouble and is too well known to need any further description here. This pain is the result of oxygen lack, which causes a buildup of self-generated, but irritating, levels of metabolic end products to take place.

Mention of the lungs leads us conveniently into the next comfort-related bodily function: the act of breathing. As with all the autonomous bodily activities, we take pulmonary comfort completely for granted until something goes wrong with the system, then we quickly realise its importance to our even tenor of life. Be honest with yourself. Did you realise that you were breathing before I brought the fact to your attention? It's only when we try to run quickly up the stairs, or when we're exposed to some choking gas, that the ability of the lungs to help in maintaining comfort is truly appreciated.

In theory, their action is very simple, as can be seen by reference to Figure 2, where lung function is illustrated in schematic form. When the rib cage is opened by the body's automatic muscular action, air is drawn into the lungs because the increased space creates a reduction in the internal pressure. The oxygen in the air comes into contact with blood in the capillary cells of the lung walls and is chemically attached to the haemoglobin in the blood. A reaction takes place, forcing the carbon dioxide present in the blood out of solution, and this gas is passed into the main part of the lungs, from where it is exhaled. When the blood reaches the cells of the body where it is to be used, the reverse reaction takes place, oxygen is removed from the blood, and carbon dioxide dissolves, where it is pumped back to the lungs for disposal. Thus, the regular action of breathing is a necessary part of the circulatory system and must be maintained in order to give the blood its vital oxygen if life is to continue.

When there is any risk of failure, the body again gives warning. If the problem is minor, as is the case when you try to run after a disappearing bus, breathlessness forces a halt to the athletic activity until the lungs' primary function of supplying an adequate amount of

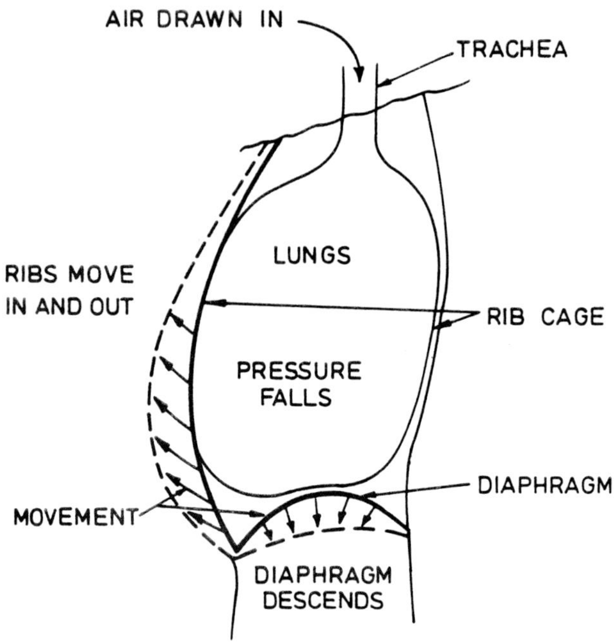

Figure 2. Diagram of the mechanical action of the lungs.

oxygen to the blood is satisfied. Where the risk is more serious (as it is when food or water is drawn by mistake into the windpipe), then a choking fit takes place. The food and air passages inside the body cross, as seen in Figure 3, so that some means of directing the right thing into the right hole is necessary. During breathing, air drawn through the nose is forced past the pharynx (the valve preventing access of liquid or solid food to the lungs). When swallowing is taking place, the larynx moves upwards and the epiglottis curves downwards so as to block the trachea (or airway) to prevent food from entering it. It's when this theoretically perfect system goes slightly wrong that trouble begins. When choking occurs, the food or liquid being swallowed pushes its way past the larynx and blocks the air passage. All eating activity is immediately brought to a halt until the obstruction is removed by an extended spell of coughing. Anybody who's ever inadvertently opened an incautious mouth in a swimming pool will be familiar with the inability to do anything

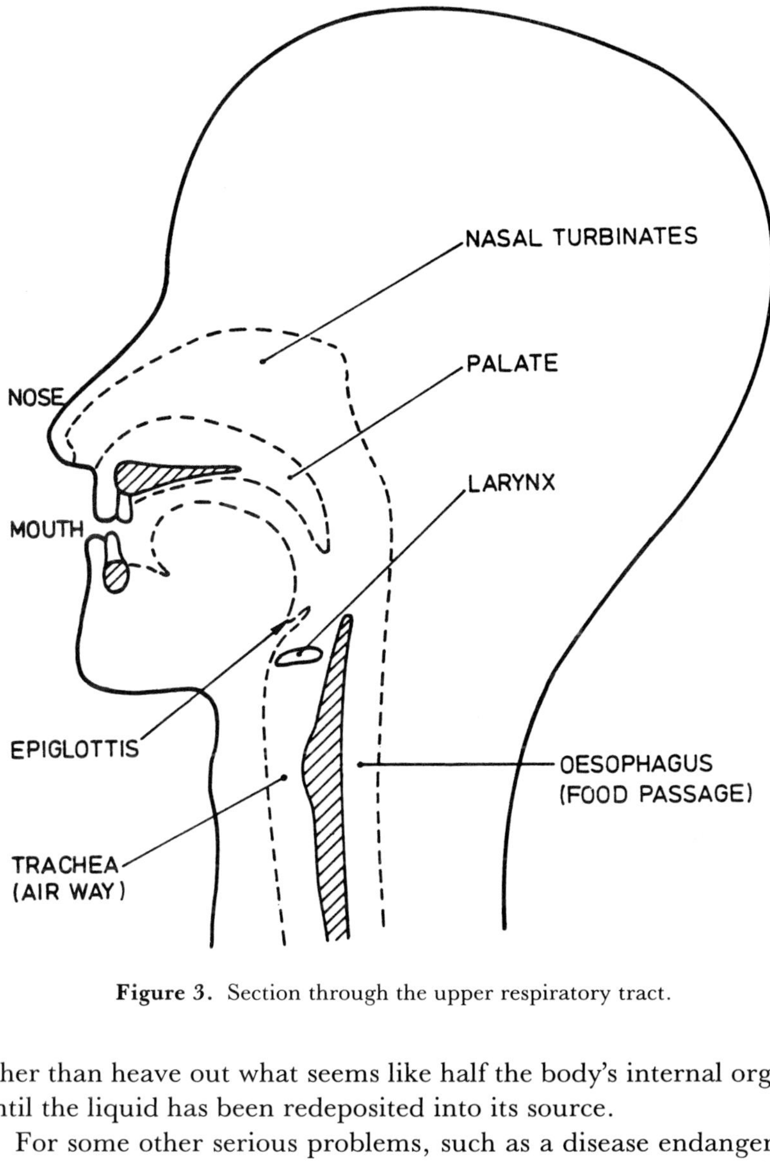

Figure 3. Section through the upper respiratory tract.

other than heave out what seems like half the body's internal organs until the liquid has been redeposited into its source.

For some other serious problems, such as a disease endangering life, the lungs may also become a site of pain, though most of the really serious lung disease is painless. As an alternative possibility, blood may be coughed up when a critical situation has developed, and the pulmonary diseases of asthma, pneumonia, bronchitis, pleurisy, byssinosis, tuberculosis or lung cancer may then not leave the victim in much doubt about their presence. Unfortunately, a great deal of painless but very severe illness does occur; in these circumstances, pain, if it comes, usually comes too late to be of any use as a warning. Even less critical problems, such as an allergenic reaction or the simple act of breathing cold air on a frosty morning, make their presence known in an unmistakeable way. Air is normally warmed to 37° C and saturated, by the upper respiratory tract, before being passed into the lungs, but a rapid intake allowing this safety process to be bypassed can produce quite a severe pain in the chest and can cause potentially harmful crusty secretions to be formed within the body.

The digestive system, illustrated in Figure 4, must also carry a major share of the responsibility for keeping us comfortable. It consists of a collection of specialised organs designed to handle the process of converting the food we eat into a form usable by the body. The digestive tract, a hollow tube about five metres in length, runs from the food-input end (the lips) to the point at which waste products leave the body (the anus). It includes also the mouth, the pharynx, the esophagus, the stomach and the small and large intestines. A point often overlooked is the fact that both ends of this tube are open to the air, so that it forms an interface between the internal and external body environments in the same way as does the skin. This interface, also like the skin, allows some substances to pass through its surface but keeps others out. As we'll see eventually, this can sometimes be a source of gastric problems, bringing with them severe discomfort.

Before looking at these problems, we should examine briefly the process of digestion itself. Apart from the digestive tract, there are additional body units, outside the tract but connected to it by ducts, which take part in the process of food conversion. These include the salivary glands, the pancreas and the biliary system, consisting of the liver, gallbladder and bile ducts. Food enters at the mouth, is propelled through by muscular contractions, and is subjected to the action of all these components in the right sequence. Juices secreted into the tract contain enzymes to convert the complicated food molecules into simpler ones. These are absorbed into the body by passing through the walls of the small intestine into the blood and the lymph,

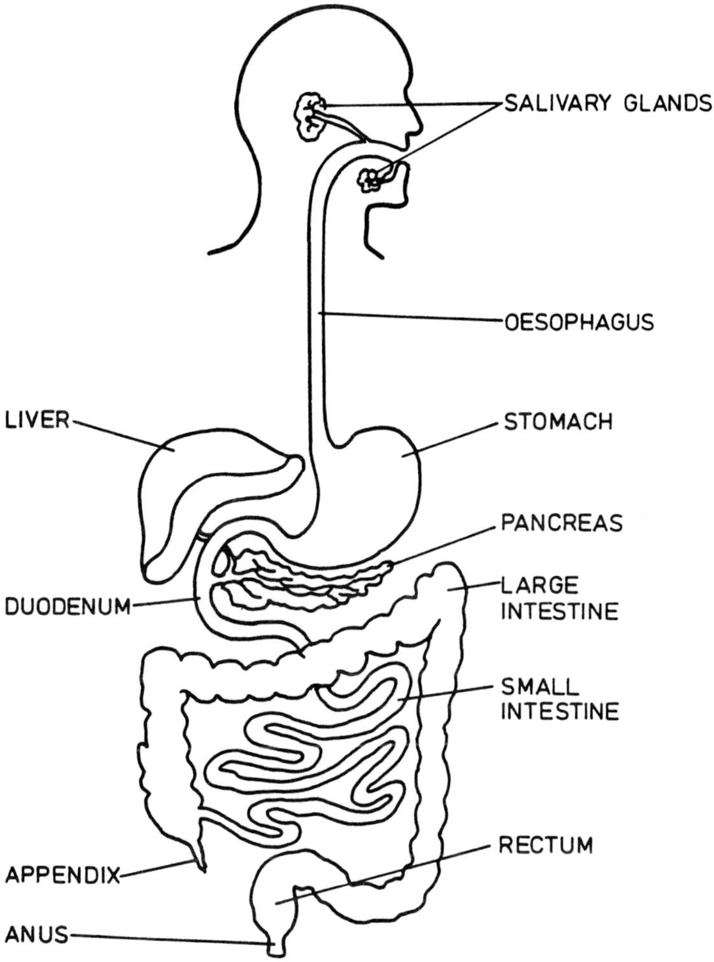

Figure 4. Components of the digestive system.

which then transport them to the body cells. At the same time, waste products are moved from the cells, via the bloodstream, to the intestine in the opposite direction and join the unabsorbed residues to be excreted eventually.

We can follow the progress of food from one end of the system to the other in our imagination. When it enters the mouth, the action of chewing reduces it to small pieces and mixes it with saliva to soften and lubricate it, thus assisting the next process, swallowing. This is a rapid transfer of the contents of the mouth to the stomach, either by gravity alone (for liquids) or by the combined action of gravity and propulsion by muscular action (for solids). Normally, valves at each end of the esophagus close between swallows to prevent excess air and too much food from entering the tract. Failure of the upper valve to close can bring on "gas pains," while failure of the lower one results in "heartburn," a transfer of the stomach contents back into the esophagus, with stimulation of pain sensors by the acidity present and with increase in muscle tension.

Once the food arrives in the stomach, its rate of movement slows considerably. The stomach expands to make room for the large influx of solids and drives the bulky matter forward by contractions, or pressure waves, at a rate of about three every minute. Enzymes and hydrochloric acid are secreted to aid in decomposing the food molecules, and the entire mixture is subjected to a churning action to bring all substances into closer contact. If the hydrochloric acid concentration is too high (and there is no general agreement on precisely how the acid is produced or its level maintained), then erosion of the stomach lining or of the duodenum (the upper part of the small intestine) can occur. When this erosion extends into or through the underlying tissue, the familiar and painful discomfort of a gastric ulcer quickly makes its presence known.

After leaving the stomach, the food passes into the small intestine to be absorbed. For this to be possible, the complex molecules in the material eaten must be completely broken down into simpler ones, and this process (begun in the mouth and stomach) now comes to completion. The pancreas, gallbladder and liver produce and transmit to the small intestine, in combination, various secretions needed to break down the resistant molecules and neutralise the excess acidity. The solid matter is moved through, by muscular action to mix and propel it, and transfer of substances to and fro across the walls of the intestine occurs continuously. If movement is too rapid (as a result of gastric irritation by toxic compounds, for example), the mixture remains wet and diarrhoea can occur. At the end of the small intestine, the residue left unabsorbed is passed into the large intestine to be consolidated before being excreted. In this part of the system, movements are much less frequent and are mainly for the purpose of increasing water absorption or transfer of the solid matter to the end region in preparation for its expulsion from the body.

Unlike virtually all other comfort-related mechanisms in the body, the process of digestion makes us aware of its presence, in no uncertain terms, even when it's working properly. The pangs of hunger are probably the most universal and common symptoms of discomfort experienced by the human race. They're integral to life in every part of the world. Sometimes, they're a permanent part of the daily existence and never disappear. For those of us who are more fortunate, they serve to remind us that the next meal is due and give us the inspiration to carry out the task of seeing that it's eaten, so ensuring a continuous supply of energy to enable the body to function at its optimum efficiency.

Hunger pangs apparently have their origin in contractions of the stomach walls, these taking place at a rate of thirty to forty per hour. They're rhythmic, each one developing to a peak and subsiding within about thirty seconds. They're faint initially, rise to a maximum and, if food isn't taken, eventually disappear by themselves after about an hour. One interesting point to note is the fact that, in the absence of a stomach, all animals (including human ones) still experience hunger drive and pangs of some kind, so that the urge to eat is indeed a dominant one. Armed with this information, you don't need to feel quite so guilty next time you raid the refrigerator at midnight.

The result of this attention to bodily needs, sadly enough, often brings about the other type of discomfort for which the digestive system is responsible. If you eat too much, you become uncomfortable. At the least, you experience a bloated sensation in the area of the navel and have to sit still for some time until it goes away. Some people may continue to experience discomfort even after the digestive process has emptied the stomach; the agonies of indigestion and heartburn after too rich a bout at the table are only too familiar to the well-fed of the world. Even at the very end of the digestive process, when a need to eliminate the waste products is felt, the discomfort of a full bowel is present to remind us of necessary duties which must be performed to keep the body functioning efficiently!

A further aspect of digestive comfort relates to the body's sensory abilities and provides a good example of how interrelated all these bodily functions can be. When an appetising smell assails the nostrils, have you ever noticed how the digestive juices begin to flow? Even the thought of a whiff of food brings a slight ache to my stomach as I sit at my typewriter, mainly because I've been working on this chapter for some time and am ready for my next meal. This is a rather unusual case where a pleasant smell (or the thought of one) can bring about a sensation of discomfort. Usually, pleasant sensory experiences increase our comfort level. Think of the way you feel when you smell a rose or the pleasure you get from one whiff of your favourite female's perfume if you happen to be male (or lesbian; I'm sufficiently broad-minded to let anybody buy my book if he, she, or it is prepared to pay for the privilege.)

We should next consider the way in which fluids affect comfort. Although hunger and thirst, strictly speaking, are entirely separate bodily functions, they are so closely related in our thinking that it makes sense to treat both of them together in a book of this kind. Thirst is set off by dryness in the throat or mouth, a dryness which can arise from lack of overall body moisture or by local changes. So, for instance, you can be thirsty if you've been out in the hot sun all day without a drink, or you can be thirsty if you've had plenty of fluids but have been breathing dust or plaster in a mad household cleaning or decorating spree.

We're not sure of the exact mechanism that produces thirst, but we think it's related to a failure of the salivary glands to supply enough moisture because they've become dehydrated. Even if these glands are removed, though, the total intake of water remains the same, but there is then a different pattern of drinking. Small amounts are taken frequently, rather than occasional large drinks. Thirst could also be due to loss of water from the overall body fluids as a whole, rather than from individual cells, because we stop drinking long before there has been time for cells to absorb the water we take in. We're fairly certain, though, that the level in the cells is controlled by the kidneys in response to water volume in the body and blood pressure changes.

Other possible reasons suggested for the onset of thirst include the stress of water depletion throughout the entire organism, a change in osmotic pressure in the blood, and cell contraction caused by loss of water from body cells to the blood. Ideally, there should be no water transfer between blood and cells at equilibrium, a state which is achieved by a balance between the amount of fluid we drink and the amount we excrete, either as urine, perspiration, or in breath. Animals, for instance, seem to take up exactly the correct amount of liquid when they drink to restore them to a fixed, and presumably optimal, body weight. If humans had the same sense, the soft drink peddlers and brewers of the world would join the ranks of the unemployed!

The thirst control centres of the brain again appear to be concentrated in the hypothalamus, that incredibly busy region of which we know a tantalisingly small amount. In this case, though, there's normally no problem of regulation. Unlike food, drink can't be stored in tissues as is the case with fat deposits. There are fat people but not wet people (in the strict sense of the word), if we exclude the temporary effects of external immersion or internal overindulgence. Extra storage of water, unless you happen to be a camel, is virtually nil in normal health. Water content, in all but the rare case of diseases such as dropsy, remains constant at about 75% of body weight. By contrast, excess food storage up to about four or five times body weight can exist in severe obesity cases.

The other senses, too, can be influential in establishing the level of comfort prevailing. The taste of a favourite flavour can bring a sense of delight to the mind, often even if only in the imagination. The sight of a mountain or of a pretty girl (with apologies for the masculine slant of these examples) can give a sense of peace or excitement, as the case may be, which is pleasant to experience. The sound of a dreamy lullaby, or of a Beethoven symphony, is potentially able to influence people similarly. The feel of a silk dress on your skin (I *knew* I would find one for the ladies eventually!), or of a piece of intricate carving under your hands, can give you a sensation of tactile comfort which pleases you whenever you remember it afterwards.

On the other hand, an unpleasant sensory experience can destroy your comfort utterly. A bad smell or taste can bring about feelings of revulsion. Picture yourself living near a pig farm or eating your least favourite food (mine are peanut butter and baby octopus in olive oil, if you're interested; separately, not in combination. I've had them both on memorable occasions and they certainly affected my comfort level drastically, I do assure you!). Is the sensation which comes to mind a pleasant one? I don't somehow think so. Remember the last time you saw something hideous — a piece of modern art, a slum area, a polluted lake, or whatever it is that fills you with revulsion — and ask yourself if you felt comfortable at the time. Bring to mind the sound of your teenager's record player wiping out decibel records or the feel of your fingernails scraping down a blackboard; you don't need me to inspire you with thoughts of sensory experiences you'll never be able to remember without a shudder shaking your frame!

Then there are the occasions where severe comfort problems are caused by sensory events. The food which causes you to be sick because it's bad, the smell which makes you retch because the gas causing it is so noxious, the light so bright that it blinds you, the sound so loud that it deafens you permanently, or the object so rough that it scrapes away your skin; these are all cases of discomfort so severe that a major protective mechanism is called into action (and, in some cases, overwhelmed). We're now getting beyond the realm of digestive or sensory comfort and into the problem of safety or survival, straying over into the area which is rightly governed by the nervous system. We should, perhaps, deal with that aspect of comfort at this stage.

The nervous system is one of the less familiar parts of the body to most nonmedical people. It consists essentially of the brain, the nerve fibres which carry messages to or from it, and the tissue cells where the brain's commands are carried out or where the stimuli giving rise to messages originate. Figure 5 shows a diagram of the nervous system, from the brain to the nerve cells in the skin which are responsible for the perception of sensation.

The brain is an extremely complex organ, and each area has a particular function to fulfill. All parts of the body are associated with their own specific representation there, but the region of most interest in the context of this section lies centrally, near the top of the organ. The sensory centre, where messages are received, and the control centre, where commands are initiated, are the parts of the brain directly associated with the peripheral nervous system. From them, the nerve fibres run down the spinal column to the point

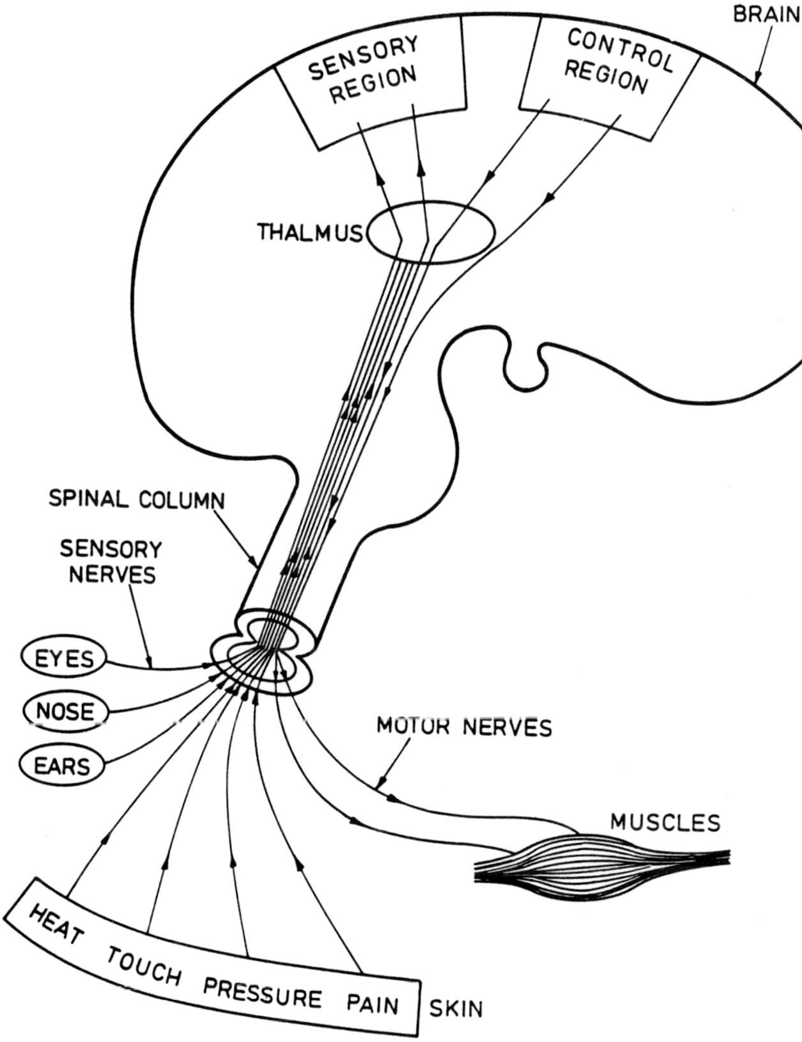

Figure 5. Schematic diagram of the nervous system.

where they must branch out to reach the part of the body served by each specific nerve fibre.

There are at least two schools of thought on the subject of tactile sensitivity. One of these says that the different sensations are perceived by stimulation of different types of cell, and the other that a different pattern of impulses is produced by the cells in conjunction when a stimulus reaches them. A compromise solution, as usual, has been suggested and proposes that a combination of these two extreme opinions can be the real manner in which cell stimulation is transmitted to the brain. That is, the receptors are specialised, in that they act as filters to block certain aspects of stimulus and pass others, so producing the different signals. Even though the precise nature of the connection between skin structure and response to stimuli is not understood though, several aspects of the mechanism can be discussed in simple terms to arrive at an understanding of tactile comfort problems. Whether the stimulus is set up directly by specific cells or by a filtering effect, we can associate specific cells with particular sensations.

The different types of sensor are located at different depths within the skin structure. Thus, for example, the contact sensors, the cells which respond to the slight touch of any other surface coming into contact with the skin, are close to the outer layers as are also the cells detecting cold and pain. The heat sensors are deeper, so we feel pain or a cold touch more easily than a hot one. The deep pressure cells, on the other hand, are buried further within the tissue, so that a blow which penetrates relatively deeply is felt twice (once as pain and once as pressure) and hurts more than we might expect. This partly explains why a scratch or slight cut doesn't at first have any effect on us. Only when the knife or abrasive object penetrates well into the skin do we notice that we're injured. A severe burn is extremely painful (and can bring about shock) because it causes damage deep within the tissues and affects many of the pain cells.

The entire tactile sense depends on this complex network of receptors in the skin. The Meissner corpuscles — egg-shaped cells near the interface between dermis (the underlying layer) and epidermis (the exterior, waterproof layer) — appear to have some function in tactile sensitivity, since they're concentrated more heavily in regions of high sensitivity. Sensitivity is measured as the closest distance of separation between two stimuli which can be detected as distinct, different ones and varies from a maximum at the finger ends and tip of the tongue to a minimum at the spinal area. It can be enhanced by training; a blind person, for instance, usually develops an amazing ability to detect minute surface features with the fingertips.

In all parts of the body, the basic mechanism of touch is the same. The nerve endings at the receptor sites are stimulated by the contact, temperature, etc., and a series of pulses is generated by each one of them in consequence. Figure 6 gives an example of the appearance of typical pulses if they're measured by the use of sensitive equipment. When a stimulus is applied at varying levels from very low to very high, nothing happens at first. There appears to be a "threshold" level below which no sensation is experienced and above which an electrical signal is suddenly sent to the brain. There's no intermediate stage; the signal is either not present at all or is there at full strength. Each pulse has a fast rise time and a slower decay, the total time of duration being about one-thousandth of a second.

As the stimulus intensity increases, pulse height, shape and time of duration remain approximately the same, but the number of impulses passing in unit time begins to increase. Eventually, a saturation level is reached and, even if the stimulus intensity continues to increase, no further change takes place in the signal amplitude, time or frequency. Higher pain levels arise not from an overstimulation of particular nerves but from the saturation stimulation of many nerves at the same time. The information contained in these pulses is sent to the brain for decoding, any variation in intensity of the sensation being characterised as a difference in frequency or amplitude of the pulse being received. The neural message, after being decoded in the brain, is recognised there as comfort or discomfort, depending on its nature, and a correction signal is sent out, if necessary, to tell the skin about the stimulus at that position.

There's still some mystery about the mechanism by which all these sensations are transmitted from the site of their detection to the brain and back, but it appears that a tiny electric current is generated by some means. There are two types of nerves, the *afferent* (or sensory) ones and the *efferent* (or motor) ones. Any triggering of cells causes the electrical signals to be sent along the afferent ones to the brain, and a command from the brain along the efferent nerves causes muscle contraction (and hence a movement). Each nerve is connected at its end to another one's end by a synapse, or junction,

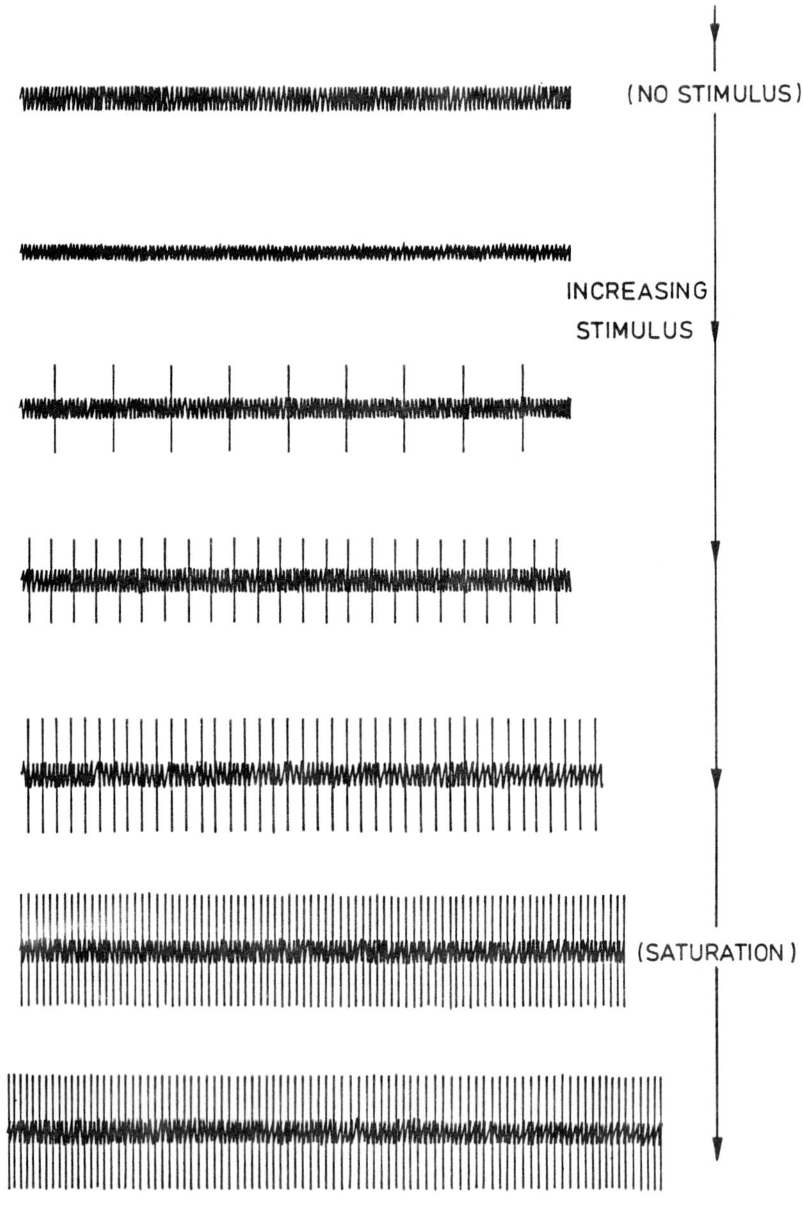

Figure 6. Typical nerve pulses.

the space between the two ends (the synaptic cleft) being about 150 Å to 1000 Å in length. Transfer of information across the gap is almost invariably chemical (and thus unidirectional) in humans but may also be electrical and reversible in some species.

Eventually, the chemical train of signal information reaches the brain, passing to the hypothalamus first and being relayed from there to specialised areas of the cortex, where a chemical reaction characteristic of the stimulus perceived is triggered. The brain then senses that reaction and interprets it as a message that something it recognises is happening at the site of the original stimulus. The comfort aspect is dependent on the number of cells stimulated and the intensity of stimulation. If there are many cells affected or if the stimulation is severe, the brain interprets the effect as discomfort, but a slight stimulation may be interpreted as a comfortable one. Thus, the gentle touch of a loved one on the side of your face will be a pleasant sensation, but a good, hearty slap will be painful and thus interpreted as discomfort.

In emergency, where an overload situation exists, a "short-circuit" mechanism is available. This is the reflex, a system that bypasses the brain. The intense stimulus causes a correction to be applied by direct contact between sensory and motor nerves within the spinal column by means of a "stored programme" incorporated into the central nervous system (CNS) by the brain. There are several thousand of these reflex actions built into the body, the most noticeable ones being the blinking action, the involuntary jerk as a knee or elbow is tapped, and the sudden move away from a hot contact.

If the stimulus is too slight, however, there may well be a different type of discomfort inflicted on the body. Imagine a feather being run gently down your spine or on the sole of your foot. There is certainly no intense force involved, nor is the feather likely to scratch you; yet the feeling is an uncomfortable one because you're tickled by the light touch. You get an uncontrollable urge to scratch the place where the feather has been; that is, you need to exert more force on the cells that have been lightly stimulated before they'll stop bothering you. This apparent anomaly, the fact that you've to increase the stress on a cell before it will return to a quiescent state, is one for which medical science has no clear answer. It may be that a cell is "sensitised" to expect a force which never arrives and remains in a state of readiness until the expected contact materialises. Alternatively, the brain may send out messages to determine whether a breakdown in the communication has occurred when the expected stimulus doesn't arrive. We just don't know and must await the results of a great deal more careful research before we can make any inspired guesses about the matter!

Now we come to the body mechanism which is probably most influential in determining our level of comfort on a day-to-day basis. This is the thermoregulatory system, the complex arrangement by which our body ensures that it remains at a temperature at which it can continue to operate satisfactorily. This mechanism is involuntary, highly efficient, and extremely sensitive. It functions ceaselessly, sometimes tangibly, but generally without our being aware of any part of its operation. At all times, we're likely to lose or gain heat to or from our environment, but the body must remain at a fairly constant temperature (of about 37° C) if we're to continue living in a normal state. If it falls too far below this value, we go into hypothermia and may eventually die of cold. If it rises too much, fever sets in and we can die of heat stroke. The permissible limits vary with individuals, depending on such factors as age, health, metabolic rate and so on, but if the core of the body is held at a temperature more than about five degrees higher or lower than the normal value for an extended time, the effect is likely to be fatal.

Now, how does the body keep itself within these very narrow limits when the outside temperature fluctuates wildly away from the optimum value? Temperatures ranging from 58° C to -90° C, according to my copy of the *Guinness Book of World Records,* have been experienced at various parts of this planet, and, if they've been measured, it seems reasonable to assume that people were there at the time. Simple arithmetic tells us that the poor old body must either get rid of about twenty excess degrees or find almost a hundred and thirty of them from nowhere, on a continuing basis, if its owner hopes to live to tell the tale. Before considering how it achieves this apparent miracle on a daily basis, we should first examine the question of where the heat comes from and where it goes.

Heat reaches the body in several ways, the most obvious of them being by direct transfer from the sun. Radiation from our nearby star is arriving constantly on earth and is the reason why life as we know it can exist here. When the radiation strikes anything, including a human being, it's converted into heat and absorbed, tending to raise the temperature of the object struck. The energy impinging on the body may be short wave (as received directly from the sun) or it may first have been absorbed and reradiated by the surroundings, in which case it's long wave in nature. Heat is merely the tangible symptom of increased molecular motion; the molecules of the heated body must move faster in order to use up the energy supplied to them, and this increased motion manifests itself as a rise in temperature. Thus, when the energy from the sun reaches our bodies, they get hotter.

The next way in which we receive heat is by the indirect route, via the food we eat. Energy from the sun is converted by the plant into food energy, a chemical form, and is then absorbed by the human being, either directly by eating the plant or by way of another animal which has first done so. The metabolic process inside the digestive system breaks down the food and converts the energy locked up in it to a form in which it can be used by the human being. The entire life process depends on this energy. It supplements the energy obtained directly from the sun and fuels the various activities taking place within the body. Basal energy (the amount needed to keep the cells alive while you lie basking in the sun without moving a muscle, for instance), the energy used by the muscles to do work, either in normal movement or in the violence of physical activity, and that needed for the involuntary muscular movement involved in shivering, heartbeat, breathing or the metabolic activity itself, are all part of the energy-consumption process.

All these forms of energy are mechanical ones, but other kinds (such as chemical, osmotic, electrical or thermal) are also produced within the body. All of them, however, depend ultimately on the chemical energy absorbed from food. The energy transformation diagram shown in Figure 7 illustrates how the body makes use of the different types and demonstrates that all the energy eventually ends up as heat. It's particularly important to note that none of the reactions are reversible, so that thermal energy (or heat) must ultimately be the sole form of energy remaining as the end product. Thus, inevitably, the ingestion of any energy tends to increase body temperature. We can control the rate of generation of heat, to some extent, by the rate at which the conversion is carried out. If, for example, we exercise vigorously, the rate is accelerated, as you'll notice if you put this book down for five minutes and sprint around the block a couple of times. The effort will make you feel hot, because you've increased the rate of energy conversion by your unbelievably foolish impulse.

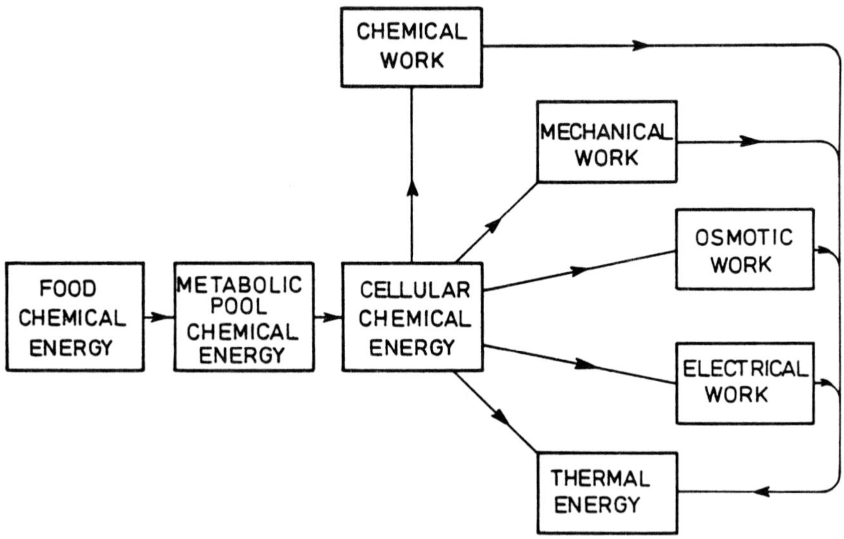

Figure 7. Energy transformations in the body.

The body has one more way of generating heat, but this one, unlike the others, isn't occurring all the time. The mechanism is the act of shivering, reserved for emergency use when the sun, the metabolic rate, and the act of exercising are unable, even in combination, to provide enough heat to maintain body temperature at the desired level. When these three modes of heating, which are operating all of the time, need extra help, the skeletal system tries to come to the rescue. The muscles all begin to tremble violently, thus reminding their owner that more exercise is needed to keep the core temperature at a healthy level. At the same time, heat is generated by the movement, so increasing temperature in the process. The mechanism is truly an emergency one, and, unless the warning is heeded and some other means of providing energy found, survival may be at risk. A newborn infant, in particular, is at a disadvantage, because this form of assistance isn't available until a baby is several weeks old. For that reason, among others, infants require special protection from the cold in order for them to survive in winter conditions.

So far, we've considered only the ways in which the body can gain heat. In permanent Arctic conditions, nothing more would be needed, but we've seen that there are occasions when the environment is too hot for survival without some form of cooling arrangement. In the human body, this can take place in one of four ways. Three of them are shared by all other objects, and the other is peculiar to only a few animals. The commonplace ones are convection, conduction, and radiation, listed in order of their importance in the context of cooling human bodies. Convection is one result of the fact that any gas, when heated, becomes less dense and is thus forced upwards by the pressure of cooler (and hence denser) molecules of the gas. Air near the body is heated by contact with it, becomes warmer, and is therefore forced to rise. As it goes upwards, it flows over the surface of the body, picking up more heat, and, when it's finally lost by rising upwards away from the body in the region of the head, it takes this heat with it. In the meantime, of course, more cold air has arrived to take its place and the process can continue indefinitely, even after the body temperature has decreased to a satisfactory level. The effect can be increased dramatically in windchill conditions, when a cold stream of air produces forced convection and causes more rapid movement of air (and hence heat exchange).

Conduction is the direct transfer of heat between two surfaces in contact with each other. If, for example, you place your hand on a piece of metal on a chilly day, the surface feels cold. The metal, which is a good conductor, is withdrawing heat from your hand and taking it away, creating a deficiency of energy which your brain interprets as a cold feeling. This kind of heat transfer can only occur when there is a temperature difference between the two surfaces and is reversible. That is, heat will flow from the body to anything in contact with it as long as the body is hotter than the other surface, but the direction of flow will be reversed if the body becomes colder than the surface. The mechanism is thus not a particularly helpful one, since we're usually too hot when the temperature around us is hotter than the body temperature and vice versa.

Radiation, too, isn't very useful for cooling human beings. It's the same process by which energy is transferred from the sun to the earth and, like conduction, depends on a temperature difference for its existence. In this case, though, the temperature difference has to be large for useful transfer to take place. Few of us can claim to be as hot as the sun, so, in consequence, we radiate heat extremely slowly, if at all. Any loss which does take place is of a long-wave form of energy and is increased by the bodily mechanism known as vasodilation, which will be considered in due course.

Now we come to the body's most useful cooling device, one which isn't found in exactly the same form in any other species or elsewhere: the peculiar phenomenon of perspiration. The term usually brings to mind those television commercials reminding us to avoid giving offence (and hence discomfort!) to other people by remembering our daily dose of deodorant, but it goes much further than this in practice. Whether the manufacturers of Brand X realise it or not, the body is producing perspiration all the time, not just when we get emotionally or physically active. Moisture from within the body is exuded through the skin and, when it reaches the outer surface, evaporates. For most of the time, we're unaware of this process and it's then referred to as "insensible perspiration." This is thought to be a result of the reaction undergone by the cells of the skin as they age. In their decay, they form keratin and give off water which permeates through the skin at a rate of about 500 g a day.

On those occasions when we're aware of moisture production, the process ("sensible perspiration") is an indication that we're becoming too hot. Sensible perspiration is generated by the sweat glands, which total two million and are distributed unevenly throughout the body surface. It's mainly water with a small amount of salt (which decreases as the body becomes acclimatised to higher temperatures). The glands are triggered by heat, under the controlling influence of the hypothalamus, and can produce a water loss of up to 2000 grams each hour. Emotional stress can also trigger the production of sweat, which is then usually confined to the palms and the soles of the feet. In either of these two states, the production of sweat increases for a very good reason: the need to cool the body.

As the liquid forms at the surface of the skin, it begins to evaporate. This evaporation, just like the evaporation of water when you boil the kettle, needs heat to convert the liquid into vapour. The only source of heat around is the body, so heat (technically known as *latent heat of evaporation*) is withdrawn from it, providing a cooling effect which tends to correct the bodily problem. (It can, of course, bring on other problems which belong to the realm of the manufacturers of Brand X; surface bacteria of the skin can thrive on the lipid content of sweat and produce the familiar odour so beloved of avid TV addicts.)

We're now ready to look in some detail at the entire operation of producing and disposing of heat within the body, which is maintained by the thermoregulatory system. This mechanism ensures that we stay alive, and at a comfortable temperature, when the environmental temperature fluctuates to a value well outside our body's natural tolerance. Without it, we'd merely adopt the temperature of the environment, just as reptiles or other cold-blooded species do, and be forced to hibernate in winter when all cell activity stopped. The production or removal of heat takes place automatically, without any need of action on our part (except for the occasional wipe of a bedewed brow so that we can see, maybe), at the correct rate for the conditions prevailing. The way in which this control is achieved is one of those miracles of nature that are so simple, yet so effective.

The temperature of the body in deep central regions like the heart, brain, lungs or abdominal organs (the "core temperature") is about 37° C. These large vital organs comprise about 8 percent to 10 percent of body weight but produce 72 percent of body heat during rest. When exercise is taking place (and when blood flow to the skin is increased drastically), their contribution drops to about 25 percent, so it's clear that there must be a very large difference between core and skin temperatures at rest. Skin temperature is very variable, with a mean value of about 33° C, the highest values occurring at places where vital organs or large muscle masses are covered by the particular skin area. The skin covering low muscle regions (e.g., the hands and feet) is at the lowest temperature and suffers the greatest heat loss. In cold conditions, the volume at core temperature tends to shrink, and a temperature gradient (of range

37° C to 20° C or colder) can occur between core and extremities. At high ambient temperature, the core temperature extends to virtually all regions of the body.

If the body core temperature becomes too high, the blood vessels near the surface of the body expand by a process known as vasodilation. In consequence, blood flows more quickly to the skin and can be cooled more rapidly, thus getting rid of the body's surplus heat faster. As blood flows to and from the extremities, the returning blood is warmed by the blood flowing outwards, so tending to equalise the temperature and preventing the incoming blood from cooling the core too much.

When the core temperature drops, the converse reaction takes place. The blood vessels near the body surface are shrunk or even (in emergency) closed completely, so keeping blood away from the cold exterior where it would carry precious body heat away to the environment. If the core temperature doesn't rise reasonably quickly, so that this vasoconstriction persists for an appreciable time, the cells in the region of blood withdrawal are starved and die off. Eventually, frostbite (the common name for this end result) takes place and a portion of dead body tissue (usually at an extremity, furthest away from the body core) may have to be amputated in order to prevent the spread of gangrene. The entire process is modified considerably by the level of activity, since this will govern the rate at which blood flows around the system and to the extremities. Cooling can also be reduced by the clothing worn, as we'll see later, since the use of insulating layers (to cut down heat loss by conduction) or tight regions (to diminish convection cooling) at the ankle, wrist or waist will preserve heat in the body considerably. As before, the counter-current warming of returning blood prevents undue cooling of the core.

Note that it's the body core temperature which controls the system; if the deep regions of the body are at a satisfactory temperature level, then what happens on the exterior is, to the brain, relatively unimportant. This is why, when you wrap warmly in winter, your body won't try to prevent frostbite in ears or fingers when these bits become cold. In theory (though I must confess I've never tried it), you're less likely to suffer from frostbite if you take off all your clothes on a freezing day than you do if you keep some of them on. By stripping, you force the thermoregulatory system to work harder, thus providing more protection for the entire body, including the bits which are exposed when you're fully dressed. One hears tales of blonde Swedish girls going out into the snow with their boy friends and beginning to, . . . but I digress. With my luck, frostbite would attack instantly and at the worst possible place!

The thermoregulatory system, then, has three major types of components: the sensory receptors (or bits that tell the brain that the temperature has fallen somewhere in the body), the central integrator/controller which initiates the correction (and which is located in the hypothalamus), and the effector organs (or bits that apply the corrections ordered by the brain). The operation of the system is illustrated schematically in Figure 8. The receptors supply information to the controller, which compares temperature with a reference value (the set point) and supplies a correction signal if needed. Signals are sent to the central nervous system, in the form of nerve impulses, from the thermoreceptors (cold or warm) just below the skin and deep in the body.

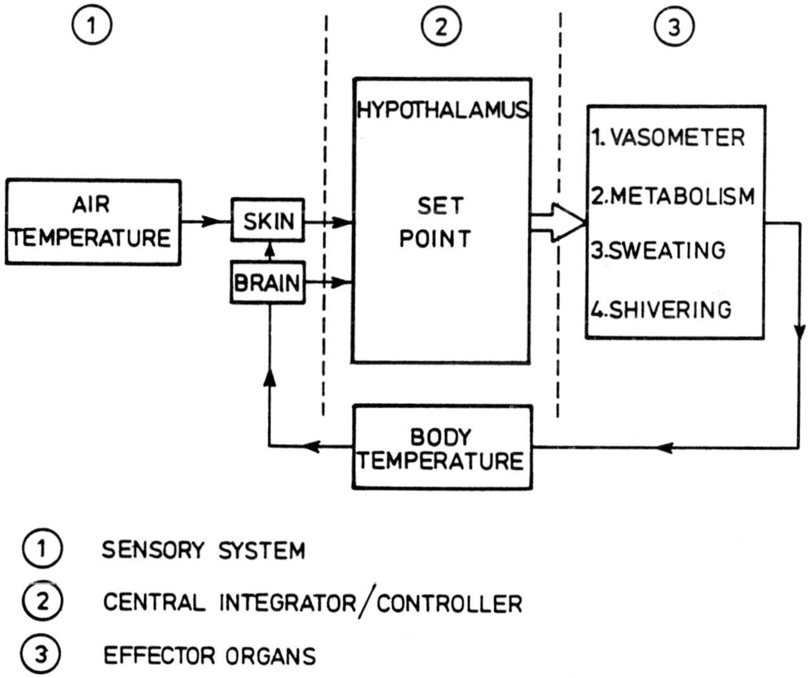

Figure 8. Diagram illustrating the operation of the thermoregulatory system.

Additional receptors of a different type are also present in the hypothalamus itself; if a cold receptor there is stimulated (i.e. by heating), anti-rise responses (such as sweating, vasodilation or panting) are initiated. Stimulus of a warm receptor, conversely, initiates anti-drop responses such as shivering, vasoconstriction or piloerection (the familiar goosebumps, which are intended to erect hairs and thus increase insulation but which are now not as useful as in the times when man was hairier than he is today!). The responses may also be of major muscle groups (the act of turning on the air conditioner or reaching for a garment, for example) as well as of the smaller ones characterised by shivering or sweating. Acclimatisation can be achieved in about six to nine days, after which an increase in sweating or in blood flow to the skin becomes automatically programmed into the body in a hot climate. The converse situation can occur, for example, in tribes which don't use protective clothing in cold climates.

Temperature regulation is affected by various problems. Skin disorders (such as sunburn or prickly heat, two of the most familiar ones) can interfere with the process, which explains all those peculiar sensations you may recall after baking at a moderate heat on the beach in the Riviera or Florida last year; so also can fainting, a temporary interruption of blood flow to the brain as a result of excessive vasodilation or other cause, and heat stroke, a rise in brain temperature above the critical value of 41° C, which can lead to unconsciousness, coma, convulsions or even death, and which is brought about by a failure of the sweating mechanism for some reason. Heat exhaustion (which is sometimes confused with heat stroke) and cramp are caused by a salt imbalance within the body and can also influence the thermoregulatory process adversely.

Mention of the topic of skin disorders brings me to the question of skin hygiene. The exterior of the human body is laden with bacteria accumulated from external sources. If they're allowed to remain undisturbed and if conditions are right, these bacteria will begin to grow. They'll make their presence known initially by an odour, the same one which brings millions of dollars annually into the pockets of the manufacturers of deodorants. At this stage, they're relatively harmless (though they may produce some discomfort either in the person on whom they're lodged or on nearby people), and the modern North American fascination with body odours is a little too paranoid to be healthy. It's not shared to any great extent to Europe, and I've never seen any evidence to show that the Europeans are any less healthy than North Americans. Indeed, the evidence seems to point if anything in the opposite direction. Perhaps the occupants of the North American continent are so worried about how they smell that they drive themselves into an unnecessarily early grave in the process!

Beyond this stage, however, there is cause for concern. If the bacteria are allowed to thrive (on humid, warm skin particularly) for an extended period, then danger to health can arise. Infection, a penetration of the skin by foreign matter which causes inflammation, may be evident. If it gains a good hold, then the blood vessels dilate, causing redness, as a result of the increased flow which the body produces in order to get more white cells to the site to fight the infection. Pain arises because of the increased pressure and the presence of chemicals acting on the nerve ends. The end result, if the body's defences are unable to cope with a massive penetration of bacteria quickly, can be such diseases as boils, abscesses, ulcers, or similar unpleasant consequences. These may require medical treatment and may ultimately prove fatal if neglect continues.

The final bit of the human organism that I want to discuss in this chapter is the skeletomuscular system. The skeleton is the framework on which all the rest of the bits are hung, and the muscles are the things that make it possible to move the framework and contents. We'll be discussing this subject in more detail in the next chapter, when we consider the physical aspects of bodily comfort, but there is one small point that can be made at this stage. When a muscle is working, a chemical reaction is taking place inside the cells of which it's composed. In strenuous exercise, oxygen can't be delivered quickly enough to the muscles to bring about complete oxidation of glucose (delivered also from the bloodstream) to carbon dioxide and water. Instead, a different set of reactions, as illustrated schematically in Figure 9, occurs.

As glucose concentration build up, the equilibrium shifts to compensate, and the end result is the production of lactic acid. When the work load is light, the blood can dispose of this compound without difficulty, but if a severe load is placed on a muscle for an extended time, the rate of production of the acid is too great for the body to

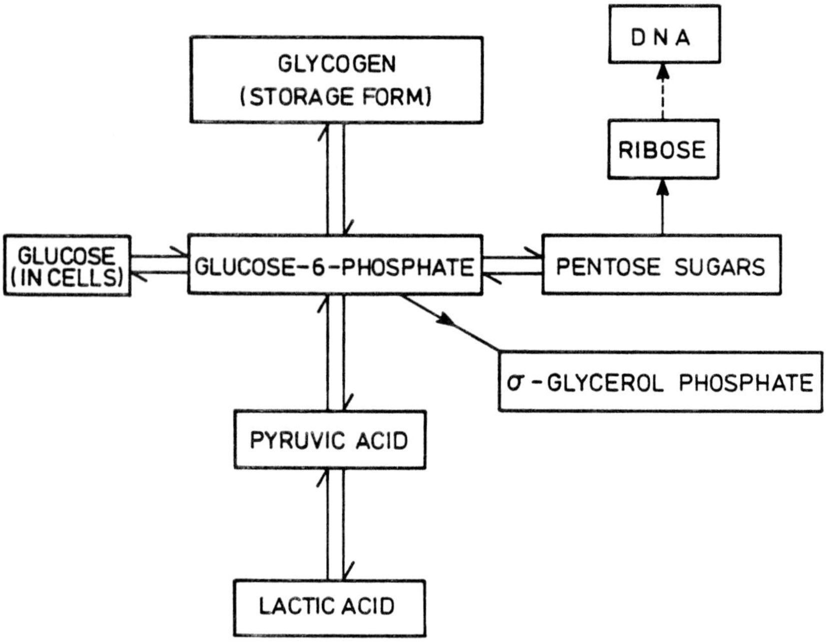

Figure 9. Muscle chemistry reactions.

dispose of it. The concentration builds up, and the sensory system (as usual) passes on this information to the brain in the form of pain. Hence, we sense the fact that a muscle has been worked too hard by the ache or cramp we experience in it. We feel inclined to rest until it can recover its normal state and tend to let it relax so that the buildup of lactic acid can be eliminated by the body's chemical reactions. If the muscle has been stressed beyond its elastic limit, we say that it's torn, or sprained, and the pain becomes both more severe and more long-lasting because of the actual stretching of nerve fibres. Until the problem is healed, its presence is forced on our attention by this pain, so that we allow time for recovery before stressing the muscle again.

These, then, are the most important physiological aspects of comfort. It's important to remember, and I can't stress the point too strongly, that they're all intricately related. It's impossible to experience discomfort in any area without one or other additional parts of the body being affected. For a human being to be truly comfortable, there must be a complete absence of discomfort in every part of his being. The discomfort need not be physiological in its primary origin. It can arise from purely mental causes or can be forced on our body by the external physical effects which form the material for the next chapter.

CHAPTER 5

PHYSICAL ASPECTS OF COMFORT

WE'VE ALREADY SEEN that the physiology of the body can have a great deal of influence on human comfort, as well as mentioning the fact that physical effects are usually, if not always, manifested by the changes they bring about in the level of psychological or physiological comfort. It's now time to examine these effects a little more closely. They can be divided into two distinct types. The first of them is the actual physical cause, within the body, of the changes produced, and the second type is the physical properties of objects within, or external to, the body.

Within the body, there are fundamentally four functions in which physical changes occur to bring about an alteration in the level of comfort, and these correspond to the four senses of sight, hearing, taste or smell. (I'm assuming that the sense of touch is stimulated only by external forces, a postulate which I think is reasonable in the vast majority of cases. There are a very few instances where an argument to the contrary can be made, but they generally have an external component if you look carefully enough at the situation.) Let's consider each of these in turn, examining first the parts of the body associated with a particular sense and then the changes occurring when comfort level is altered.

Vision depends on the optical system of the eye, illustrated schematically in Figure 10. Light reflected from a visible object passes through the pupil and is focused by the lens to form an image on the retina. This is a complex layer of tissue composed largely of nerve cells. Two types of cell are important in vision and are called (from their approximate shapes) rods and cones. Directly behind the pupil, at the centre of the field of view, the sensory layer consists entirely of cone-shaped cells, which have the ability to detect colour. As distance from the centre of the field of view increases, the incidence of cones steadily decreases, and successively more rods become present. These tend to detect only light intensity, not colour, but are much more sensitive to the presence of light than are the cones. You can easily demonstrate this fact for yourself by looking up at the stars at night. Find a bit of sky that looks completely black, and look at it carefully. Try to find, at the edge of your peripheral vision, a faint star. When you've located one, transfer your vision centrally to it and you'll find that you can no longer spot it. Look to one side of it again and it will reappear. Thus, the "corner of your eye" is a much more sensitive area than the front, which explains how the teacher could always catch you misbehaving as a child even though he wasn't looking your way.

On the other hand, the eye sees with greatest clarity only in the

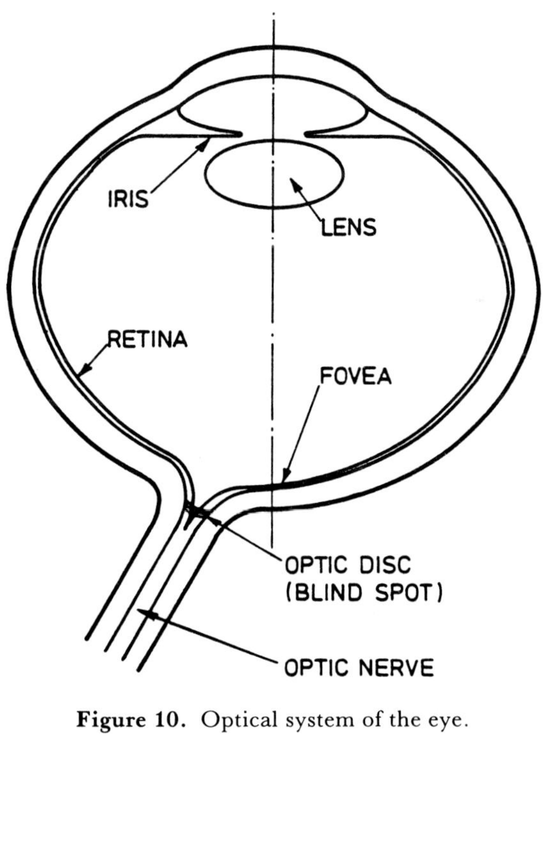

Figure 10. Optical system of the eye.

central region of the field and gives an indistinct image at the periphery. Thus, the star (or the child) is only seen as a blur, not sharply, which accounts for the one time you were falsely accused. The cones are connected individually to nerve fibres, whereas the rods are grouped together first, each group being connected to a single fibre which the cells share. When light rays fall on a nerve cell, they set up a minute electric current in the cell, and the cumulative sum of these currents is collected at the optic nerve, to which all the nerve cells are connected, for transmission to the brain, where it gets decoded. At night, an increase in sensitivity occurs as a result of a chemical reaction in which a compound called "visual purple" is bleached. This substance is only regenerated by the rods in conditions of darkness, which accounts for the fact that it takes some time to become accustomed to seeing on a dark night.

If a bright light shines in your eyes, protective mechanisms come into effect. The most obvious ones are the contraction of the pupils and the ability of the eyelids to screen out rays by folding over the eye. If there's insufficient time for the eyelids to close, or if they're held forcibly open when a bright light is directed into the eye, damage to the cells is likely to occur. The damage may be temporary or permanent, depending on the intensity and time of exposure, and blindness can ensue if a substantial number of the cells are destroyed in this way.

One effect which is often noticed is that of watering, a mechanism by which the body washes the eyes to remove any foreign matter likely to harm the tender membranes. This rinsing action takes place every time we blink, though (as usual) it's only when discomfort is extreme that we become aware of it. Another comfort-related factor in the sight sense is the problem of eyestrain, an ache in the muscles controlling eye movement. If these muscles have been overworked (by prolonged concentration on the print of a book in dim light, for example), they become tired, like any other muscle, and use pain to bring the fact to our attention. Persistent eyestrain pain is an indication that the muscles have become permanently damaged and can no longer fulfill their duty of eye-focusing without the aid of spectacles to assist them in this task.

The sense of hearing is next on our list of body mechanisms influenced by physical aspects of comfort. A diagram showing the principal parts of the ear used in hearing is shown as Figure 11. Any sound which we hear consists of a series of vibrations of the air molecules, and the disturbance of these vibrations moves rapidly through the air. Eventually, a part of the wave reaches the outer ear, the funnel-shaped bit stuck on the side of the head, which is designed to gather sounds and channel them into the middle ear. Between the two lies the eardrum, a membrane which can be set into vibration when the sound waves strike it. As it vibrates, it causes three small bones in the inner ear to move in synchronism, and their movement is then transmitted in turn to the inner ear, to which they're acoustically connected. The vibrations are conveyed by the fluid in the oval window to the cochlea and then to the tiny nerve endings connected to the auditory sensing mechanism of the brain.

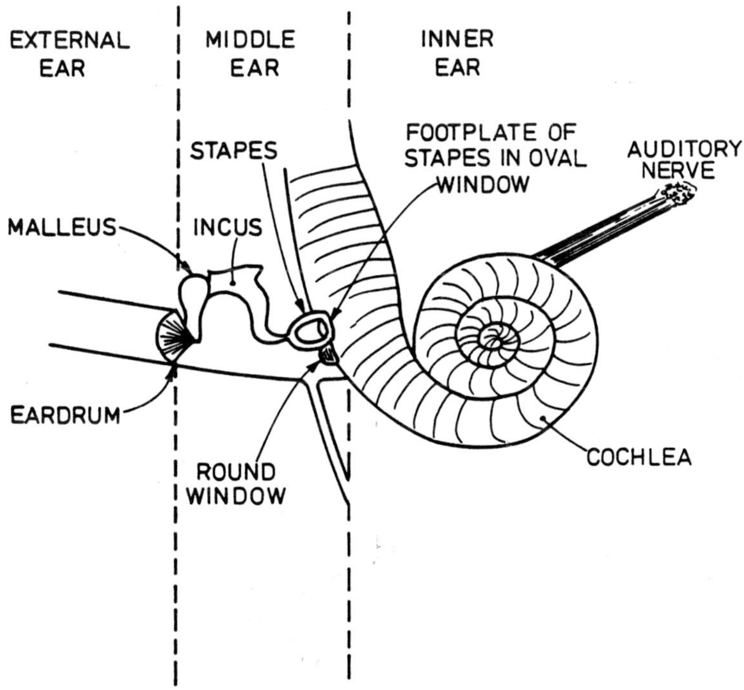

Figure 11. Diagram illustrating the structure of the ear.

The range of hearing, like that of vision, differs from one person to another. A healthy young human being (unless a devotee of loud rock music, of course) can detect sounds ranging in frequency from about 2 Hz (though tonality can't be perceived below about 20 Hz) to about 25 KHz, where the hertz (Hz) is one cycle per second. The range of amplitudes detectable is from about zero dB(A) to 140 dB(A) of the sound pressure, where the decibel, one dB(A), is a defined standard unit of sound intensity. The sensitivity of this amazing organ can be appreciated when you recall (as I'm sure you all do) that 0 dB(A) is a vibration with the same amplitude as that of a hydrogen molecule!

The ability of the ear to detect sound depends on both the frequency and the amplitude of the sound source. Figure 12 shows the minimum audible field (MAF), or the lowest possible sound which can be heard by a normal, healthy human being, as a function of fre-

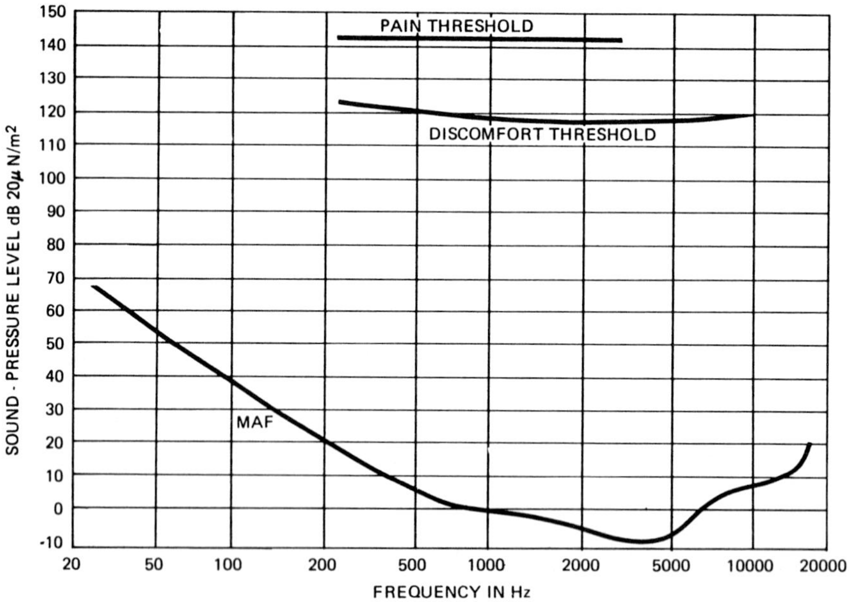

Figure 12. Frequency dependence of hearing sensitivity. (Reproduced by kind permission of General Radio).

quency. As can be seen, the best sensitivity is in the range 2000 Hz to 5000 Hz, though sensitivity is good from 200 Hz to 15 kHz. In the middle of the range, where the ear is most sensitive, a change in frequency of about 0.03 percent is detectable and a reduction in intensity of about one decibel can be recognised.

The upper limit to frequency is the speed at which the sensory hairs can respond to rapid vibrations of the air as a result of inertia. At the other end of the frequency scale, it's the damping-out effect of fluid movements around the sensing components in the ear that stops ultra-low sounds from being heard (at higher frequencies, the vibrations are too rapid to allow this damping movement to take place). The ability to discriminate, within the brain, between genuine and spurious signals limits the lowest intensity detectable. Sensitivity is governed by the minimum stimulus that can be detected over the intrinsic random "noise" of the hearing system's neural triggering in the absence of any external stimulation. The range of movement of the mechanical parts of the ear causes pain and, hence, limits the upper level of noise tolerated, when vibration becomes too violent. Figure 12 also shows two curves representing the level at which discomfort and pain, respectively, become noticed. In the limit, the maximum tolerance is governed by the tendency of the cochlea to suffer damage if vibrations are too intense.

Exposure to the upper levels of intensity for more than a short time can cause deafness of a temporary (where relief is obtained) or permanent (where exposure is prolonged or repeated on a regular basis) nature, and it's now illegal in most Western countries to subject human beings to noise levels above about 85 to 90 dB(A) on a routine basis. At high levels of ambient noise, workers must be removed from the environment after a defined period of exposure, and maximum exposure times for a working day of 8 hours are strictly observed. Typical industrial derating values, showing the permissible working time in a noisy environment as a function of noise level, are given in Table I. With age, or exposure to loud noise, the range decreases at a rate dependent on many factors.

TABLE I

PERMISSIBLE WORK EXPOSURE TO HIGH INTENSITY NOISE LEVELS

(typical limits imposed by legislation)

Sound Level db(A)	Permitted exposure time (hours)
below 90	No restriction
90 to 92	6
92 to 96	4
96 to 98	3
98 to 100	2
100 to 102	1.5
102 to 105	1.0
105 to 110	0.5
110 to 115	0.25
above 115	No exposure permitted

Unlike the eye, the ear has no means of automatic protection when exposure to a stimulus too great for normal operation occurs. There are no earlids, so no safeguards against the sustained effects of loud noise exist. If a teenager chooses to play the record player at a high level of intensity or to attend rock concerts on a regular basis, he or she is very likely to develop deafness over a period of time. Workers in noisy industrial situations, too, are candidates for the early ear trumpet. Unfortunately, the risk hasn't always been appreciated in the past, and the onset of deafness is an insidious change which isn't noticed until the damage is done. Figure 13 illustrates typical progressive hearing loss occurring over a period of time by showing the hearing sensitivity changes (as a function of frequency) at different ages for a worker exposed continually to sound at high levels throughout his working life. This type of progressive deafness is caused either by fusion of the bones of the inner ear or, more usually, by damage to the nerves carrying sound to the brain as a result of overloading.

Hearing discomfort, then, can take two forms. The listener may be unable to hear properly, either because the source is too faint or as a result of deafness. The effort to distinguish the words or sounds then places a mental strain on him, and the comfort is of a psychological kind. Alternatively, discomfort may be the result of too loud a noise, in which case there may be both physical and mental aspects. Pain may be experienced as a result of the intensity, and the victim

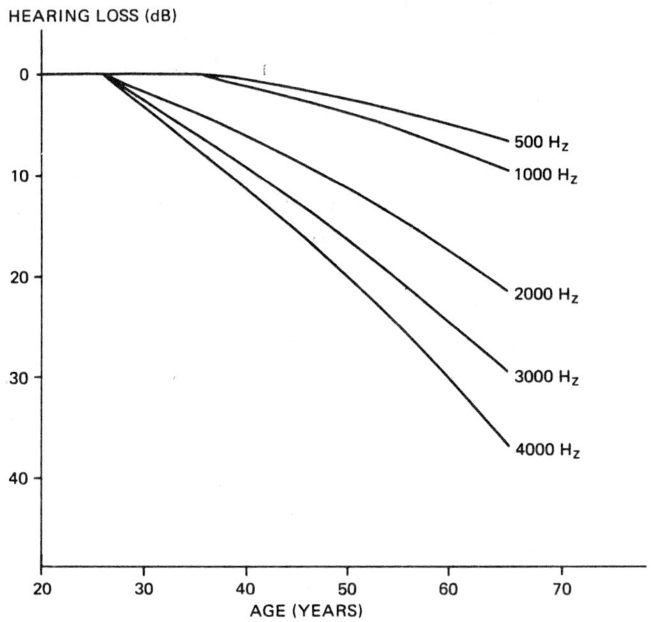

Figure 13. Long-term hearing loss curves.

will certainly be distressed mentally in consequence. Nobody likes to be forced to listen to an unpleasant sound, especially at high intensity, and if the intensity level is really high, as in an explosion, actual physical damage may result. In addition, the act of listening to something unpleasant in nature, even at moderate levels, is itself an uncomfortable sensation. I recall one summer in Austria, where I was working in the holidays from my university, the weekly ritual of killing a pig at about four in the morning. It invariably woke me, even though the task was done at a far enough distance for the squeals not to be too loud. By some feat of courage, I managed to eat the pork served in various guises for the rest of the week, but I dreaded Tuesday, the day set aside for the slaughter!

The ear has one other comfort-related function not connected with the hearing aspect. Inside each inner ear there are three semicircular canals, containing within them a gelatinous fluid known as endolymph, which give information about the direction of gravity and the movement of the head in space. Any motion of this fluid produces agitation in the hair cells of the semicircular canals so that a movement of the head is thereby sensed. If a hair is bent in one direction, the rate of firing of its receptor cell increases, while it decreases if the hair is bent in the other direction. With the head in the normal position, little activity is produced in the receptors, but a tilt brings about changes in output in three dimension. Motion is detected also from information conveyed from the joints and from pressure receptors in the soles of the feet, as well as from stimulation of the nerves in the ear, as a result of acceleration.

All these signals are integrated by the central nervous system, and correction is an automatic function of the reflex mechanism. If correction is needed, information about the orientation of the head is computed and instructions are sent to various muscles in an attempt to keep the body in equilibrium. If there is a problem anywhere in this mechanism, the balance is upset and the person suffering from the disorder may tend to topple over. Intense or abnormal stimulation produces vertigo (a turning sensation), nausea, and vomiting as experienced in motion sickness. The most usual reasons for a malfunction of this kind are related to diseases of the inner ear, though a head injury or tumour may influence the sense of balance on occasion. Drugs or other toxic substances can also have an effect on the balance mechanism as may easily be verified on a Saturday night outside the local hostelry (though the doubtful pleasures of getting drunk involve the cerebellum as well as the cochlea).

The sense of taste, too, has some part in establishing our comfort level. The tongue, the organ playing the major role in detecting a taste, is about three inches long, contains bundles of muscles that allow it to move about in many directions, and is covered with many tiny projections (papillae), some of which are surrounded by trenches. In each of these trenches, lining the walls, are 200-300 taste buds, a single one of these being illustrated in Figure 14. A taste bud is shaped like a flask, includes about 15 to 60 elongated cells, and contains at its base an opening through which nerve fibres enter. The cells, acting as receptors, are of different types (or possibly of a common type at different stages of development) and have microvilli, fine finger like projections extending into the taste pore opening. Molecules which stimulate taste receptors are assumed to drop into the taste pores and interact with the microvilli. Nerve fibres enter through a membrane at the base of the cells and carry the impulses generated by the detection of a given taste to the brain where, as usual, they're analysed. The papillae in different areas of the mouth detect different aspects of taste. Those near the back of the tongue transmit the sensation of bitterness, those at the top of the tongue sweetness, and those on the sides of the tongue recognise saltiness and sourness. The only necessary condition for satisfactory operation of the taste buds is that the substance detected must be in solution, and the salivary glands moisten the entire area for this purpose, among others.

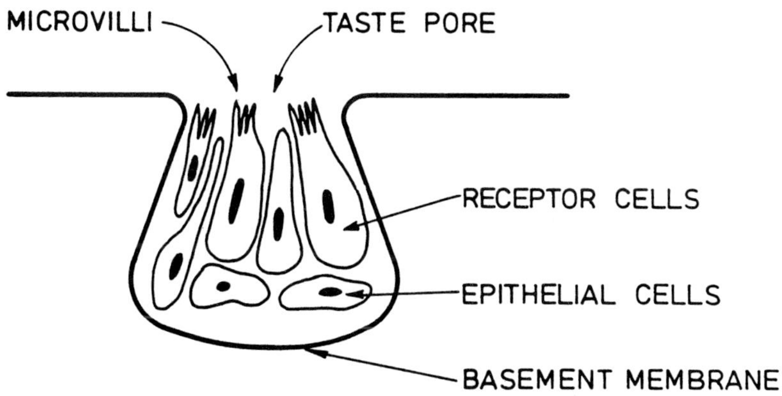

Figure 14. Cross-section of a taste bud.

Taste discomfort is a difficult quantity to define. In the first case, taste is very subjective. A food which may be very pleasing to one person is anathema to another. Hot curry is a case in point; either you like it or you're normal, like me. It's a little incredible to realise that the vast differences in taste all stem from the balance of the four fundamental components — bitterness, sweetness, sourness and saltiness. Different tongue areas are indeed highly specific in sensing the three last types, but bitterness can apparently be detected everywhere, possibly by a slightly different mechanism. Obviously, there are environmental, and possibly hereditary, factors which come into play in determining the individual preferences we all have. There are, however, some universal limits beyond which nobody would accept a given taste as pleasant. If a substance is too far outside an ill-defined area of any one of the four components, it tends to be rejected universally.

Unfortunately, the science of tastology, or whatever name it acquires when the jargon-mongers get at it, has never been developed to any quantitative level, so that we can't establish the location of these limits. Nevertheless, there is general agreement that if something is too bitter, it's uncomfortable to keep it in the mouth and we follow the natural inclination to spit it out. Similarly, sourness and saltiness, though not usually as unacceptable as bitterness, can become so great that comfort is impossible while the sensations remain in the mouth. Over-sweetness, though usually regarded as less uncomfortable than an excess of the other three sensations, can be too cloying to be acceptable. Although no work has been possible on the magnitude of taste sensations, it seems reasonable to assume that the sensory mechanism can accept only a particular quantity of stimulus of any one type, generating signals of discomfort when this acceptable level is exceeded. Whether the levels are different in the four cases, or whether the difference in apparent tolerance of the four components is illusory, can never be established until some means of measuring taste perception quantitatively has been devised. I'm prepared to stick out my neck and forecast that this breakthrough won't occur for some considerable time, if ever.

Now for the sense of smell. This one is even more tenuous than that of taste. The nose is more sensitive in detecting faint odours than even the finest instrument. It's also capable of discriminating between smells of very slight difference, a fact which keeps such people as perfume and wine testers employed in ensuring that their product meets the demand of human beings the world over. We know that the nose, illustrated in Figure 15, is lined at the back of the nasal cavity with a sensitive area where smell is detected. It contains three types of cell (base, support and receptor), and odorous molecules interact at the cilia, or hairs, of a receptor. The mechanism by which this reaction produces a stimulus is virtually unknown, but there is some suggestion that an odour correlates with the shape of the molecule and its physical "fit" into sites on the surface membrane of each cell. When odour molecules strike the olfactory target, a cell fires, giving a pattern of pulses characteristic of the particular molecular shape. These pulses are then carried to the brain, as usual, and recognised there.

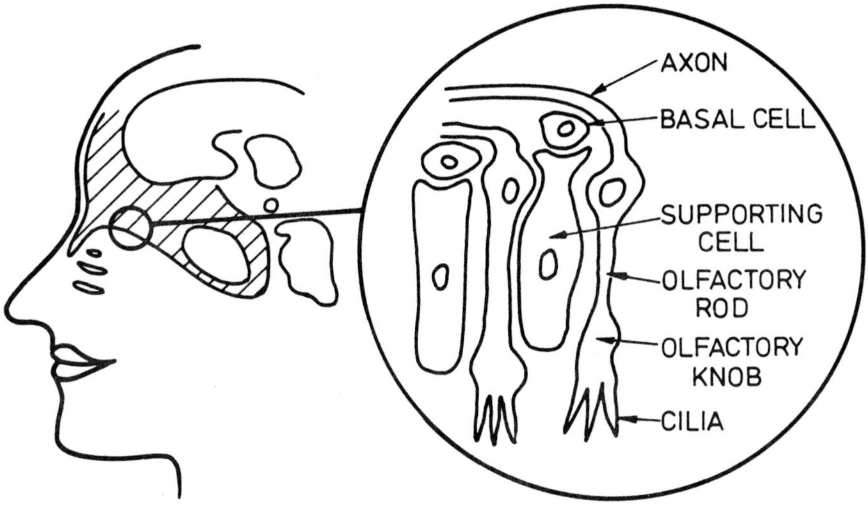

Figure 15. Schematic diagram of the nose.

According to this theory, there are seven different types of molecules which give rise to seven different types of odour. These odours are listed as those resembling camphor, musk, flowers, peppermint, or ether, together with the smells described as pungent and putrid. For the latter two, it's thought that olfaction involves a chemical reaction in which molecules of these two odours combine with specific cells in the nose, or with chemicals in those cells, in the first part of the process which ultimately generates the impulses sent to the brain for detection.

Since we don't have any quantitative way of describing smell, it's correspondingly difficult, once again, to establish the reasons for odour discomfort. We assume that, as with taste, there are limits of molecular excitation which are universally accepted as defining the comfort threshold, but again we can't discover any real meaning for them. Clearly, odour discomfort and comfort are important for human beings; you only need to visit the cosmetic department of the local stores on a busy day to be made aware of the fascination which perfumes have for many of our species.

There is, too, general agreement that the senses of taste and smell are related in some mysterious way as you can easily prove by eating an appetising meal or by trying to distinguish between an apple and an onion with a clothes pin on your nose. Also, a saturation "shutter" of some kind exists in both cases. When you've been exposed to a given taste or smell stimulus for a short time, it ceases to be registered on your brain, and the same thing is true if you've gradually become accustomed to the presence of the stimulus. This is in many ways fortunate; the farms of the world wouldn't exist for long if the human brain continued to be as badly affected as it is on first acquaintance by, say, the smell of pig manure. The sense of smell is the quickest one to tire in human beings, but is also the quickest one to regain its full abilities, a fact which bodes ill for those who commute from the countryside and return to the manure odours nightly! Discomfort, then, arises from a sudden increase in the concentration of an odour, rather than from the intensity of the odour itself. Whether there is a different sensitivity to the seven classes of smell we don't know, and, as with taste, I don't predict any speedy change in the situation.

These, then, are the physical aspects of comfort which are related to the body's senses. Now we turn to the ones which are governed by other aspects of the human being or by factors external to the organism. First comes the comfort depending on the skeleton and muscles of the body. The skeleton is the framework of bones or other semi-rigid structures which supports the soft tissues of the body and which provides leverage for the muscles when they do work. In humans, the skull, spinal column and ribs form the axial skeleton, to which are attached the appendicular skeleton (the pelvic and pectoral girdles) and the appendages (the arms, legs, their associated bits, and odd bones in the head, like teeth or ear parts). To this frame, at the appropriate points, are fastened the muscles and ligaments of the body. Muscles are bundles of fibres which contract in response to instructions from the brain to cause bones to move. Ligaments, on the other hand, don't change in length during normal use but merely hold their burden in a stable position. They're firmer fibrous tissues which provide support for the joints or organs of the body, so that they can move freely as necessary without falling out of place.

Muscles only show two responses to stimulation: by shortening and by an application of force. When muscles function, therefore, they do so by contracting in straight lines. A muscle can't push or bend but only pull. When it pulls, the distance between its ends diminishes, so causing the bones at each end to move nearer to each other. Those bones, at another place, are connected by a joint, so the process of reducing the distance between them produces a bend in this joint and the desired movement takes place. The action of the muscle is magnified by the leverage inherent in the system, so that a small movement of a muscle results in a large movement of a bone. Figure 16 illustrates what happens when the muscles controlling the way your arm bends are functioning. Take the diagram with you, even if only in the mind's eye, to your next party and be thankful for the efficiency with which the body's mechanical parts work!

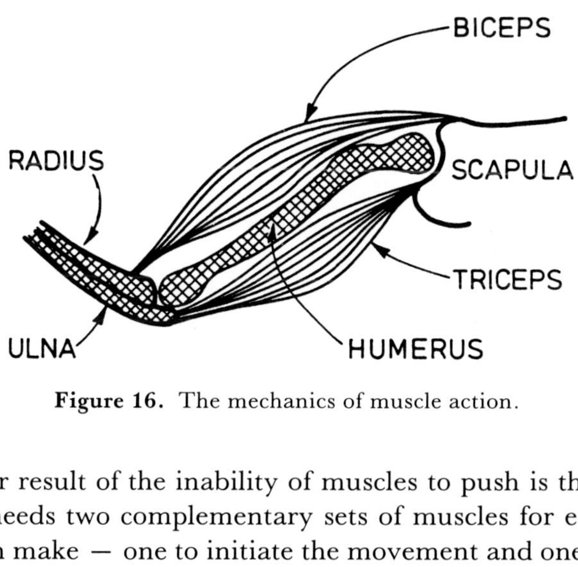

Figure 16. The mechanics of muscle action.

Another result of the inability of muscles to push is the fact that the body needs two complementary sets of muscles for each movement it can make — one to initiate the movement and one to restore the limb, etc., to its original position. The biceps, for example, tend to produce flexion of the forearm, while the triceps tend to resist flexion and are thus used in restoring the forearm to its initial state. In theory, I suppose, there should be some difference between the relative strengths of the two muscle groups in drinkers and abstainers, but I've never heard of any study of this vital subject. No doubt social science will get around to it eventually.

Discomfort in the skeletomuscular system can take several forms. The ache of a muscle given unaccustomed work to perform, as we saw in the last chapter, is chemical in origin, but there are other sources of pain which are more serious. The most common one arises from the extension of a fibre group beyond its elastic limit. If a muscle or ligament is stretched as a result of an accident or by applying an overload, some of the fibres are torn and a pain is produced. The elastic limit, the length to which the fibres can be extended under a load and still recover completely when the load is removed, is a function of the length of the muscle, its state of readiness for work (i.e. the amount of any previous exercise it has been given), and the state of health of the person owning it. An athlete is less likely to suffer from muscle strain than an unfit individual if they're both exposed to the same load (though the athlete may, of course, be more likely to experience a strain because he pushes his body to accept greater loads).

Discomfort of a different kind may arise as a result of disease, bad posture, or disuse. Rickets, a skeletal disorder caused by nutritional deficiencies, and muscular dystrophy, a crippling disease characterised by gradual wasting of muscle, are two typical diseases which can cause discomfort. Slouching, or leaning at a peculiar angle, can produce a permanent deformity which is likely to lead to discomfort in time. Working conditions in a coal mine of old, or the act of sitting in a modern car for hours on end on a daily basis, are typical examples of this. In industry, machinery controls may be placed in an awkward position, forcing the operative to adopt an unnatural stance to carry out his job. In all of these cases, redesign to correct the fault is needed. The human skeleton is resilient and has to be in order to fulfil its function, but it wasn't made to deal with that kind of constant abuse.

Effects on circulation, too, can be serious. Gravity increases pressure below the level of the heart and decreases pressure above it. If a sudden move is made from a prone to a standing position, half a litre of blood (an appreciable amount) is suddenly moved from chest to feet. This is equivalent to a fairly major haemorrhage, so there may be a faintness or darkening of vision for a few seconds as a result of the sudden drop in blood pressure and flow to the brain. Other cardiovascular effects of postural discomfort include pooling of blood in the legs (and possibly varicose veins) or swelling of tissue (of the feet during day and of the face at night, for instance).

Disuse, whether for long or short periods of time, is likely to bring about either atrophy (a reduction in size resulting from the loss of contractile muscle protein) or cramp whenever the muscle is used again. The lack of use may stem directly (as in the case of poliomyelitis, causing nerve paralysis, for example) or indirectly from illness or injury when confinement to bed or the use of a cast is necessary. It seems that a continuous connection to the central nervous system and some degree of exercise are both needed to maintain a muscle in a healthy state.

It's now time to move away from the body and examine the physical factors, related to comfort, which have an external origin. These are factors over which we've no direct control, either because they're governed by invariable physical laws or because they're related to massive changes too large for manmade control to be feasible (as in the weather, for example). The first of these physical factors is thermal insulation, probably the most important one in the context of clothing for protection against the discomfort of cold weather.

Thermal resistance may be defined qualitatively as the ability of a material to impede the transfer of heat between opposite faces of its structure. In quantitative terms, it's more usual to start with a consideration of the physical conditions governing heat flow. Consider, the example, a thin slab of material with a thickness dx and of uniform cross-sectional area A. If a *small* temperature difference dT exists between its opposing faces, and a quantity of heat dQ flows through the barrier in a time dt, then the thermal conductivity k is defined in the equation:

$$dQ/dt = -k \times A \times dT/dx$$

The negative sign indicates that heat flows towards the lower temperature, so that T decreases as x increases.

Thus, k is the rate of flow of heat through unit area of the material per unit temperature of gradient and is measured in units of watts per metre Kelvin. For any flow to occur at all, it's necessary that there should be a temperature difference between the two faces, and it's the function of any insulation to maintain the temperature difference, as completely as possible, with no change. In the absence of insulation, heat flow would equalise the two temperatures, so it's clear that a higher temperature difference between two regions (other things being equal) will be evidence of better insulation. The heat flow may occur by convection, conduction or radiation, as already discussed, so a good insulator will provide a resistance to the movement of heat by all of these mechanisms.

Heat flow occurs by transfer of kinetic energy of motion as a result of impacts between molecules in a gas or liquid, but in a solid these effects are enhanced by two further ones, the vibration in the lattice structure (this is particularly important if the solid is crystalline) and the drift of free electrons. An increase in the vibration amplitude of atoms, brought about by a rise in temperature, causes the extra energy to be passed on to adjacent atoms. The process is a progressive one, so that the temperature rise spreads throughout the structure, but the change in rate of vibration is relatively slow as a result of inertia in the large atomic masses.

Thus, conductivity is low in cases where this is the major mechanism, and substances exhibiting this type of behaviour are insulating in nature. When free electrons are present, though, they can acquire a higher velocity (and therefore energy) much more rapidly because their mass is much lower. As a result, they're able to carry this energy quickly to all parts of the solid and are good conductors. Such substances are also good conductors of electricity, as you can verify by thinking of how quickly the heat of your hand is carried away when you touch a cold metal surface. The metal, a good conductor of electricity, is also a good conductor of heat and withdraws a lot of warmth from your hand as you stand there getting colder by the second.

Now, how do we set about putting all this interesting information to some practical use? In theory, conduction and convection can both be suppressed by maintaining a vacuum between the two temperature levels, but this isn't always possible. A more usual way of preventing heat flow is by interposing air spaces in the path of the flow, preferably in small "pockets" to improve efficiency. When human beings are getting cold, it's usually transmission via a gaseous route that's involved. Conduction and convection both occur by movement of molecules of warm air to a colder region, carrying the heat energy with them, so it's this movement which must be prevented.

Air spaces which are totally enclosed won't allow their contents to escape, so conduction is stopped fairly easily, but convection is more difficult to control. The air pockets within the insulation allow heat to be carried to the top of each space and to be passed on to the next pocket where the process in which warm air rises is repeated. The most effective method is by the use of "dead" air space, in which many thousands of tiny pockets are packed within a very small volume, so that the rate at which convection currents pass heat from one to another is greatly diminished. If the pores of air are totally impermeable, then heat is prevented almost completely from escaping if the cells are small enough, and this is the principle underlying most insulating materials such as clothing, interlinings, wall coverings, and similar products.

Dead air space is particularly important in situations where any degree of external air movement is likely to occur. In winter conditions, especially if a blizzard or strong wind is blowing, the effective insulation of a human being is invaluable to the preservation of comfort. If the air next to the body remains trapped there, it's gradually warmed by body heat and acts as a barrier to the flow of heat away from the body. If, on the other hand, that layer of heated air is lost, it will be replaced by another layer (cold because of the outside conditions), and the body will provide heat for it to be warmed in turn. The continual exchange of warm air for cold air in this way will obviously drain heat away from the human being in question, so that he'll gradually become colder and colder, with serious consequences in time. If, however, the air is totally trapped next to the body, there is no real possibility of major heat loss occurring. Thus, the particular bit of the body protected by this insulation will remain warm, and hence, comfortable.

The actual loss of heat by a continuous exchange in high air flow situations is known as "wind chill" (to be discussed in more detail in the next chapter) and can make a cold day much more severe than it would be in still-air conditions. The wind chill effect is normally measured as an equivalent temperature, calculated as the temperature difference which would be necessary to achieve the same effect in still air as is felt in the existing conditions. On a severe winter day, it may be as high as thirty or more Celsius degrees, a considerable difference which may mean severe danger to the person exposed to it. The situation is made worse by the presence of water in cases where the air spaces are not impermeable to the liquid, since water is a much poorer insulator than air. Let's turn now to a consideration of this aspect of physical comfort.

Water exists in three forms — as solid, liquid or gas — and all three of them have some relevance to the topic of comfort. The gaseous type — water vapour — has already been encountered in our consideration of perspiration comfort, and we can expand that discussion a little. Water vapour moves through a barrier by a process known as diffusion, the result of chance molecular collisions. The molecules of water, as they change their location, are constantly colliding with air molecules and cannoning off in all directions. If there are more water molecules on one side of a barrier than on the other, it follows that there will be more collisions involving these molecules on the side of higher concentration. The collisions will tend to cause movement to occur equally in all directions, so that there will be more water molecules moving from the high concentration side to the low concentration side than in the opposite direction. Eventually, the numbers will be the same on both sides of the barrier, as long as no other mechanism for adding or removing water vapour molecules is introduced into the system. At that point, an equilibrium will exist in which the number of water vapour molecules moving one way will be exactly balanced by the number moving in the opposite direction.

The rate at which this equilibrium is achieved is governed by the concentration of the water vapour molecules (since the more there are, the more likely it is that they will be forced through the barrier) and by the "mean free path" of the system. This is a term which can best be understood by considering what happens to a single molecule moving within a gas, as in Figure 17. At frequent invervals, the molecule will collide with other ones and, at each collision, it will change direction. A zigzag path, random in nature, will thus be followed. It can be shown by statistics that, after such movements (if they're of equal length), the average distance from the starting point will be equal to the square root of *n*. If the rate of diffusion is measured by experiment (using a coloured gas such as bromine, for instance), it's possible to carry out a fairly simple calculation to determine how long each "step" between collisions must be, and this value, a measure of the distance travelled in a straight line by any particular gas molecule before it hits another molecule and hence changes direction, is its mean free path.

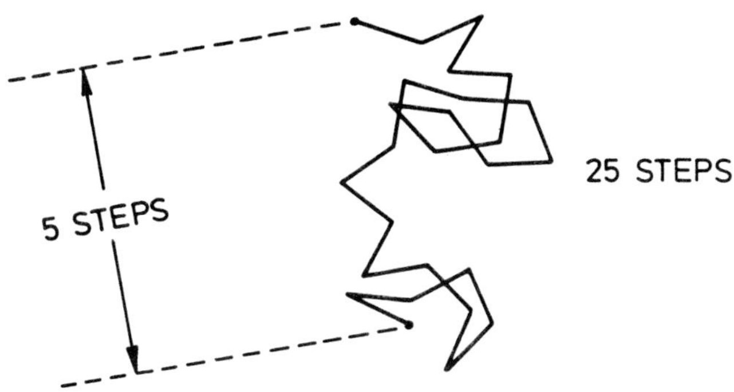

Figure 17. Mean free path of a gas molecule.

The magnitude of the mean free path will depend on pressure, since a lower pressure means fewer molecules in a given volume, and so fewer opportunities for a collision to take place. It will also depend on temperature, since molecules move faster at a higher temperature (and are thus likely to collide more often). In the particular case where water vapour, produced by the body in perspiration, is passing through clothing, there is the additional complication of fabric structure. The length of the mean free path will be influenced by the pore size of the barrier, since the water vapour molecules will collide with the molecules of the sides of the pores through which they're travelling as well as with air molecules. The smaller the pores, the more collisions there will be in a given time, the longer it will take for equilibrium to be established, and the more the perspiration will build up within the garment.

The speed at which the water moves through the barrier is an approximate estimate of its rate of diffusion, and the ability of the barrier to transmit the vapour is a function of its moisture vapour permeability. The latter term is an important one in comfort, especially of clothing, since it has a very great bearing on perspiration comfort for obvious reasons. Liquid water, too, is important in the context of comfort, as anybody caught in a rainstorm will agree. Liquid water porosity, like moisture vapour permeability, is governed by pore size, but in a different way. The movement of liquid water through a barrier which isn't completely impermeable is influenced by several factors, the most important of which is a physical property known as surface tension. All liquids possess this property to some extent, though it varies in magnitude. It is best understood by considering what happens when some water is spilt on to a horizontal surface, as in Figure 18, where the shapes taken up by a single drop of water, a larger quantity of water, and water containing some detergent are shown schematically.

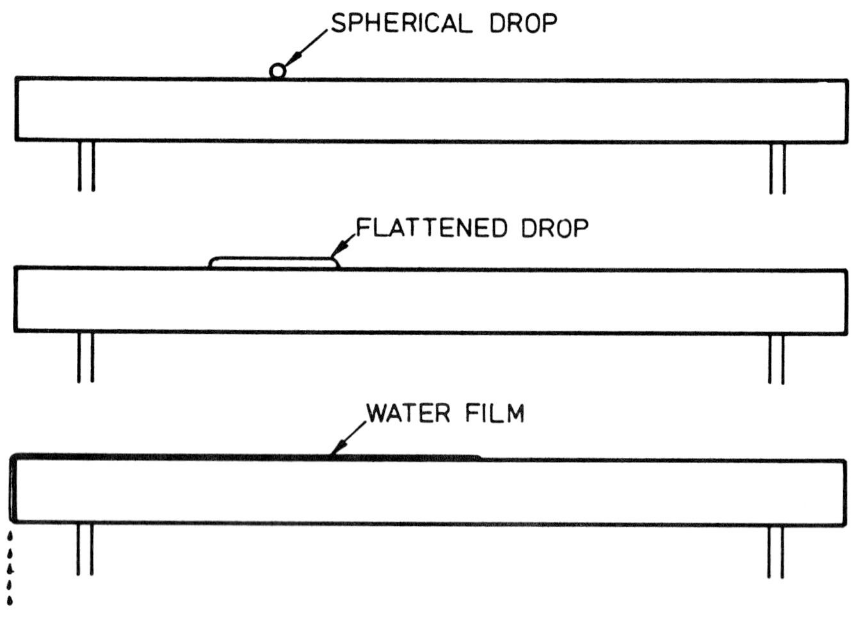

Figure 18. Surface tension in split water.

If the surface is very clean and the water is pure, the single drop will be almost spherical in shape. When more water is added, the liquid will form a pool which beads on the flat top of the surface. If, however, you now put a drop of detergent into the pool of water, it will immediately flow all over the surface and form a very thin layer of wide expanse. What's happened? There's still the same quantity of water, more or less, but it now occupies a much larger area than it did before. The answer is that its surface tension has been changed. This quantity is a force which exists in all liquids and which tends to prevent them from flowing by holding the surface molecules together closely, thus giving the surface the smallest possible area.

With one drop of pure water, the force is great enough to keep the drop almost unchanged in size. If more water is spilt, then the gravitational attraction on the water molecules tends to flatten the liquid by pulling them downwards. When we put a small amount of detergent into the water, though, we change the chemical interface between the water and its surroundings. The surface tension is reduced drastically as a result, the force holding the surface molecules together decreases, and the gravitational attraction now pulls all the water molecules downwards. The best way for them to go down is to spill past all the other molecules, just as the drops of water or grains of sand used to spill down in the days when you made mud pies. As they spill down, then, they flow out from their original place and spread all over the available space.

Now, all this is very interesting, but how does it affect your comfort, you're asking, as you fight your way home through the storm? The answer is very simple, at least in principle. It's the surface tension at the boundary between the rain and your coat which determines, in part, whether the water will go through to your skin or not. If the surface tension is high, the water will bead and remain outside the fabric. If the surface tension is low, you may begin to curse at any time in the next few minutes, because you'll eventually feel the first damp tingle to let you know you're in for a soaking. A definition of surface tension which is often used identifies it as the force per unit length acting in the surface perpendicular to one side of a line in the surface. It arises because the force of attraction between molecules in the surface is diverted, when the surface is broken, towards the object which has divided the tension-creating bonds.

Two other factors have an important influence on the amount of water reaching our skin, if you're in the mood to find out about them. The most obvious one is pressure, as you'll be able to verify if the storm gets any more fierce. If a gentle drizzle of tiny raindrops is striking your coat, it may not possess the energy needed to force itself through the fabric and wet you. Once the rain begins to descend in earnest, however, with big drops falling at high speed, the force trying to drive the rain through the raincoat increases to the point where the fabric can't resist it, so the water begins to seep in.

The other factor is the property known as wicking, which is closely related to surface tension. One of the peculiarities of liquids is the fact that they tend to creep along narrow tubes, even against the force of gravity, as a result of surface tension. The force holding the surface molecules together can also attract other molecules to the surface, including the molecules of any container holding the liquid. The effect isn't very noticeable under normal conditions, but you can observe it if you put some water into a glass and examine the circle of liquid at the edge of the surface. You'll find that the liquid seems to climb slightly up the wall to form what's called a meniscus, a tiny rim of water all round the top of the liquid. Its origin can be shown fairly simply by reference to Figure 19, in which the forces existing in the liquid at a vertical wall are examined.

Any liquid surface must be at right angles to the resultant force acting on it or the liquid would flow to correct this abnormal situation. Normally, the surface is horizontal, since it's then perpendicular to the gravitational field, but near a container wall it becomes curved. There must, therefore, be a net force acting which isn't merely that of gravity. In Figure 19, the attractive force on a molecule M, resulting from the effect of nearby molecules in the surface, is shown as MX. This is the force of *cohesion* which binds the molecules to each other and causes them to behave in the way typical of all liquids. In addition, there is a force MY exerted on M by the nearby molecules of the solid container, a force of *adhesion.* The net force acting on M, therefore, is the resultant of these two, MF, acting at an angle as shown. At this region, the surface can only be perpendicular to the force if it's curved, and the nearer the molecule is to the container, the more the surface must deviate from the horizontal. (It should be noted that for mercury and other liquids in which the forces of adhesion are greater than those of cohesion, unlike in

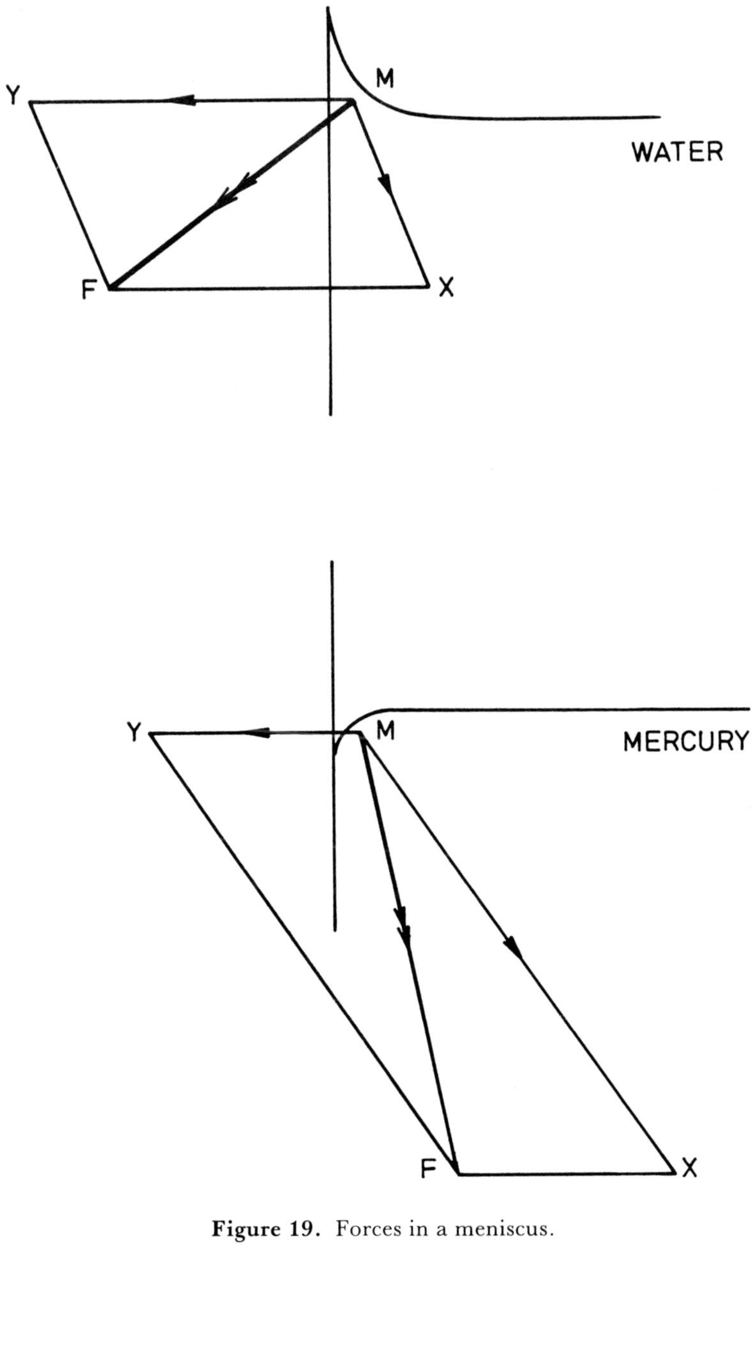

Figure 19. Forces in a meniscus.

the case of water, the meniscus will curve in the opposite sense and a slight dip will occur near the container. This can easily be seen in a mercury thermometer.)

In a normal glass vessel, the amount of water in the meniscus rim is a microscopically small proportion of the total contents, but a peculiar thing happens if you make the glass extremely narrow, as illustrated in Figure 20. When the water is placed inside a thin tube, rather than a large container, the rim can pull the surface of the liquid some distance up the tube. Surface tension in the liquid, as we've seen, exerts a force on the walls of the vessel, but an equal and opposite force is exerted by the glass on the liquid.

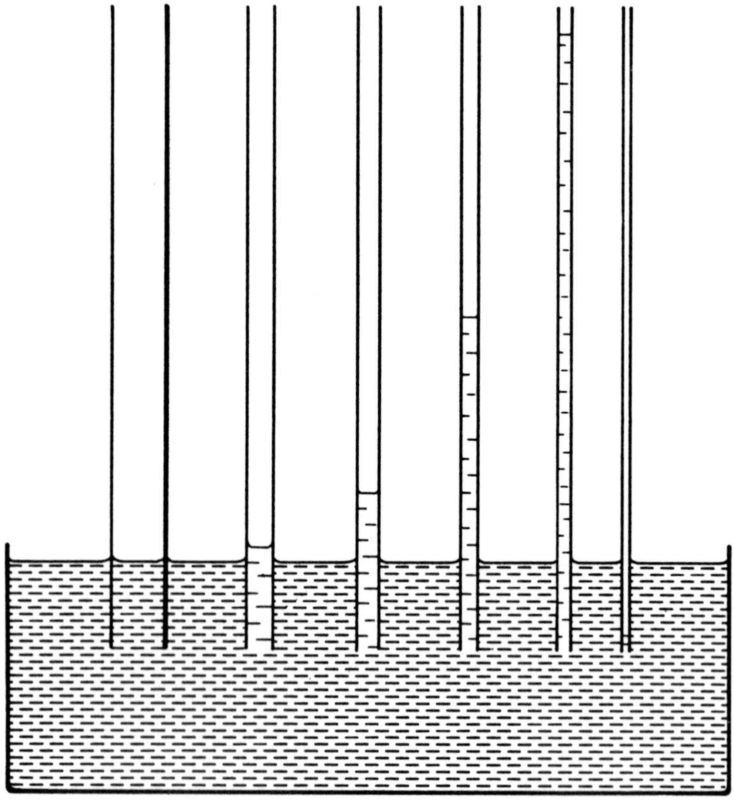

Figure 20. Capillary action in fine tubes.

This pull must have an upwards component, since it's at right angles to the (curved) surface and is only balanced by the effect of gravity on the water that has risen above the surrounding surface. Thus, the water rises until the downwards force due to the mass of this head of liquid is great enough to equal the upwards force of attraction. It follows that for finer tubes (which contain less mass per unit of height risen), the water will rise further than for larger-bore ones. As the tube diameter gets smaller, then, the distance to which the water rises increases until, for a fine capillary tube, it can reach a height of 25 centimeters or more. The limit to the distance it will climb, strangely enough, is also surface tension. As the tube becomes smaller, there comes a stage where the entrance is too small for the water molecules to enter, because they're prevented from separating by the force in the water surface. The practical result is obvious. Most coats are less than 25 centimeters thick, so the effect on comfort is a drastic one. As long as the surface tension is small enough, liquid will move through the pores of the fabric by this capillary action, and the movement becomes greater as pore size decreases.

This brings to light an important point regarding moisture comfort. When you're wearing a raincoat, you want it to do two things. It should keep the liquid water out, but it should also transmit, at the same time, moisture vapour from your body to the outside air for the disposal of perspiration. By the argument of the previous paragraph, a large pore size is beneficial for reducing capillary wicking, and large pores will also allow the free movement of water vapour molecules to take place. We've already seen, though, that molecules of liquid water can penetrate more easily when their pressure is high. Unfortunately, they can also flow more easily through large pores. Thus, a more open structure, which is of benefit for better perspiration disposal and reduced wicking, allows greater free access of water droplets through the fabric and so is undesirable. In other words, the two sets of requirements (open structure for improved moisture vapour transmission and closed structure for improved waterproofing) are mutually incompatible. As in so many aspects of comfort, we must seek a fairly unsatisfactory compromise between two diametrically opposed conditions.

Now we're ready to examine the complex problem of winter comfort in a little more detail. Let us assume that the environment is cold, at a temperature well below the freezing point of water. You're moving vigorously, or working, so that heat is being generated within your body, and there is an ample food supply stored up inside you to enable you to maintain your activity for some time to come. Perspiration is being produced, but, because the day is so cold, your coat is buttoned up against the elements. The perspiration reaches the outer layer of your clothing system, meets the cold air near the external surface, and immediately condenses to liquid water. If the day is cold enough, ice may form, but in either case the end result is the same. A cold surface has been produced within your garment system. When the next water vapour molecules arrive, then, they'll condense at this surface and form more liquid even further into the clothing array. We may express the situation by saying that the plane of condensation is moving further into the clothing system, as illustrated in Figure 21. Eventually, the liquid layer will reach the skin, either directly by continued condensation or as a result of wicking from other areas where liquid water is already present. At this stage, the first sensation of discomfort is noticeable.

There is, however, yet another problem in the attempt to maintain comfort. Now that the garments are waterlogged, the dead air space no longer exists, and thermal insulation has decreased drastically (possibly by a factor of 100 or more). Thus, body heat is used up much more rapidly, and you start to feel very cold. The situation is worst if, after a heavy spell of work, you stop for a rest. As soon as your activity level decreases, your body stops producing as much heat, so the drain caused by the combination of reduced insulation and heat lost in evaporating the water from your garments becomes much more severe. Add to that a breeze producing wind chill, and serious trouble may well be in store. At the very least, you can expect to suffer severe discomfort, and there may well be a risk of frostbite or worse.

Frostbite is a direct result of the body's automatic mechanism of self-preservation. When there is a risk that body temperature may drop to a dangerously low level, steps are taken (as already discussed) to ensure that the core temperature remains at the correct value for as long as possible. One of these steps is the closing of blood vessels near the surface of the skin, cutting off the blood supply

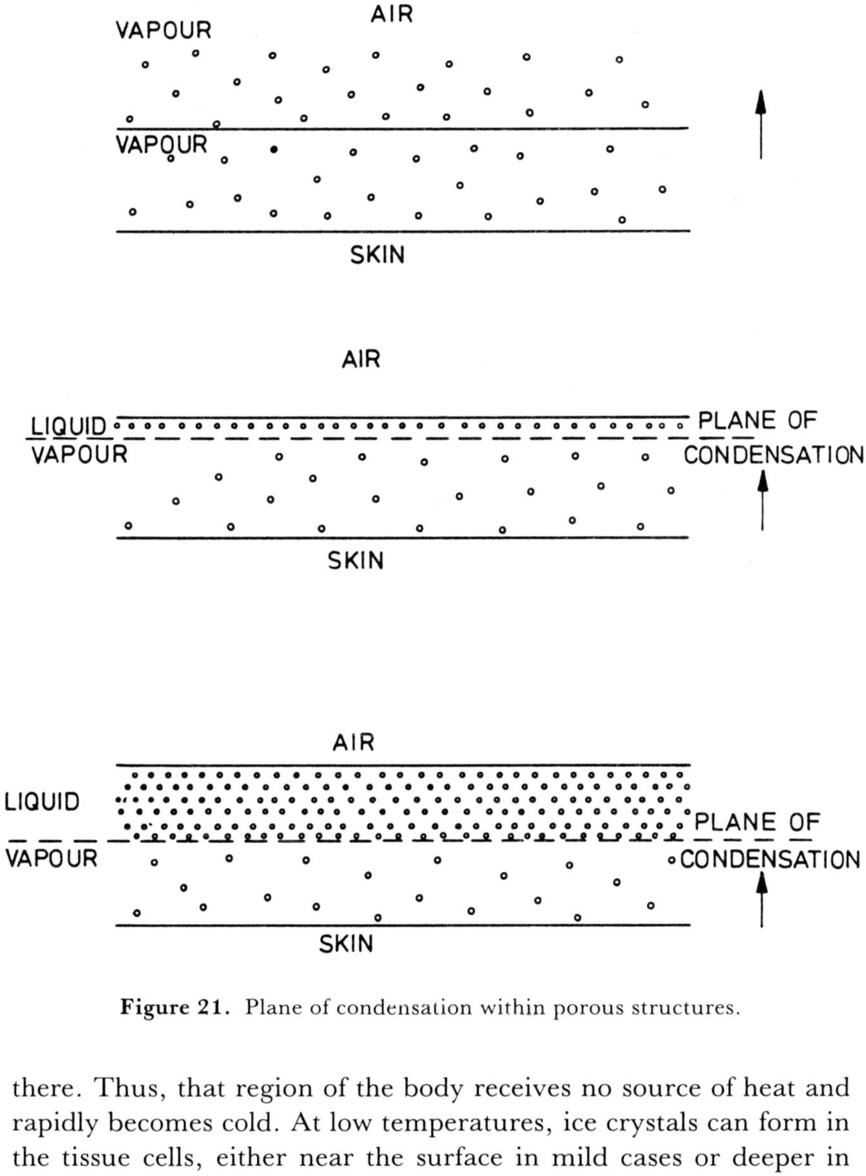

Figure 21. Plane of condensation within porous structures.

there. Thus, that region of the body receives no source of heat and rapidly becomes cold. At low temperatures, ice crystals can form in the tissue cells, either near the surface in mild cases or deeper in severe ones. The anaesthetic effect of the cooling makes the onset of frostbite unnoticeable, and if it's not detected and treated quickly, complications can set in. A redness in the affected area is succeeded in turn by the formation of blisters, tissue death, chilblains, clots in the blood vessels and, in the most severe cases, the onset of gangrene.

This scenario brings out the interconnection between all the different factors involved in comfort. The blood vessels, which contract in the body's defence, are a part of the cardiovascular system. The physical problem causing the danger in the first place affects the thermoregulatory system. The pain experienced when the frostbitten area begins to thaw is felt in the nervous system, and the misery resulting is psychological. The same observation is true in the converse situation, where the body is too hot. In this case, the blood vessels expand, more blood is pumped to the surface, and heat is lost more rapidly. Again, the physical conditions causing the problem affect the cardiovascular system, the thermoregulatory system and, if sunburn or sunstroke exist, pain and misery are felt. Truly, the comfort of a human being is a most complex matter, with causes or effects which occur in ways that you least expect!

CHAPTER 6

THE MEASUREMENT OF COMFORT-RELATED FACTORS

BEFORE we look at how comfort factors can be measured, it's necessary to consider in some detail the subject of measurement in general. When we want to find out how big, or long, or heavy an object is, we get out a ruler or a balance and put it to work. If, on the other hand, we need to know how important, or tired, or happy a person is, the problem becomes a little more difficult.

There is general agreement as to what constitutes length or mass (unless you happen to be fighting the adoption of the metric system) but not about what is meant by importance or happiness. We say that the two situations require objective and subjective measures, respectively. An objective measure is one carried out with the aid of instruments which don't depend on the opinion of a human being (though they may require human skill to obtain a reading). A subjective measure, conversely, is one in which the opinion of the person carrying out the measurement is the only factor of interest. Speaking generally, an objective measure is simpler to perform accurately and is likely to give results of greater precision. If you don't know the difference between precision and accuracy, I can recommend a good book which explains it in a clear and humorous way (and which, incidentally, I wrote!). Please write to me for details. My address is in the Preface.

Thus, when you want to carry out an objective measurement, you set up an experiment with the appropriate apparatus, you run the experiment, and you record the number or numbers which represent the value you're seeking. You then calculate the mean value and apply all those weird formulae you learned in statistics classes years ago, and you end up with an answer that is more or less correct, give or take a few of whatever it is the analysis tells you. When you need a subjective measurement, though, you can't do the same thing. There is no instrument to measure what somebody is thinking, and the only apparatus you have to tell you the answer is the tongue or pen of the person doing the thinking. If he doesn't want to be truthful, you've no way of proving that the answer he gives you is inaccurate (barring the use of lie detectors, which may or may not work anyway).

Added to that, there is the problem that people tend to have very different opinions about everything; you've only to ask yourself what you really think of my superb style of writing to become aware of how wrong you can sometimes be. This means that you need to ask a lot more people for an opinion before you can be as sure of the accuracy of your answer as you were in the objective measurement. Even when you get their answer, it's difficult to know how to go about analysing your results. It's easy to get a mean value of ten different weights, but not so easy to get a mean value of ten different answers to the question, "How much do you like members of the opposite sex?" I mean, how do you begin to average four "wow!" two "ugh!," two "Okay, I suppose," one "I don't know," and one "I'm not sure what the opposite sex is" replies?

Finally, there is the thorny question of how much you can actually rely on the result you get when you ask for a subjective opinion. People are notorious for their ability to tell lies on any and every occasion. Look at the divorce statistics if you don't believe me. Not only will somebody lie if he feels the truth would show him up in a bad light, but he also may not realise he is distorting facts. The very mention of the word "comfort" can make him realise that there's some comfort-related aspect, which he hadn't previously noticed, bothering him.

Then, even if this factor doesn't complicate the measurement, there is always the problem of inconsistencies. The way the subject feels can change his opinions. If he has got a cold, or a headache, or has had a fight with his wife or the boss, he may feel that something totally unrelated, like the colour of his suit, is no longer becoming to his masculinity or his personality. As a result, his comfort level deteriorates and the answer to your question (which may be totally unrelated to the way he feels or looks in his misery and his suit) gets a bias which you can't possibly be expected to take into account.

So, in measuring factors related to comfort, we must first decide whether a particular situation needs a subjective or objective method of assessment, then determine what experiment needs to be carried out and, finally, how many times we need to repeat it to be sure we have an accurate answer. In simple terms, we can usually expect to carry out measurement of psychological factors only in subjective ways, measurement of physical ones only in objective ways, and measurement of physiological ones in either subjective or objective ways, depending on the type of response we're seeking. The distinction will, I hope, become clearer as this chapter unfolds.

Psychological factors, the ones where personal idiosyncrasies can be so important, are the hardest of all to measure. Let's look first at the problem of estimating interpersonal distance. How close do you want to stand to somebody as you talk to him? Well, I don't know about you, but for me it all depends on who he is and what we're discussing. If it's somebody who either bores me to tears or whom I dislike intensely, my preferred distance approaches infinity. I can't get away fast or far enough from people like that. On the other hand, if it's a pretty girl I'm trying to . . . No, on second thought, I'm a respectable married man and I don't do that kind of thing. Besides, my wife might read a copy of this book. However, I think you get my point. If you like talking to the person, you're prepared to let him or her approach very closely indeed before you begin to back away.

The usual assumption made by behavioural scientists is that there is no inherent reason why you should either want to escape from, or become intimate with, the person to whom you're talking when the measurement is made. In other words, they don't have halitosis and you're not trying to seduce them (though how these scientists can be sure I'm not going to hazard a guess). With this assumption, the two people are observed, without being aware of the experiment's purpose or of the fact that they're being watched, and notes are made of the way in which they move. If they have the same optimum interpersonal distance, there will be little movement and the two will remain at about the same distance (which can be measured) from each other. If not, one of them (the one with the smaller, preferred distance) will close in until at his or her interpersonal distance and the other one will then move backwards to a comfortable distance in turn. Thus, the two will oscillate between their respective interpersonal distances, which can subsequently be determined.

The need for privacy is a little more difficult to establish in anything approaching quantitative terms. Again, it's not constant. There are times when each of us wants to be utterly alone and other times when isolation would drive us mad. When you've got a problem on your mind, for instance, you need to be able to get away by yourself to think it out, not be bothered by a load of friends who all want you to join in some kind of distraction. On the other hand, if you happen to be the friend out for a good time at the weekend, the last thing you want is this wet blanket who has nothing to do but brood over some stupid problem or other. It's not much use asking for an opinion, either. How many entertainment stars declare that all they want from life is to be left alone? And how soon are they on the phone if their agent takes them at their word? I'm not sure that a real measurement of the need for privacy exists or if it's a valid one when you carry it out. One approach is to observe undetected the behaviour of human beings who have the choice of entering or ignoring a conversation and then making judgments about their response to the choice. Clearly, unless they then give a truthful answer to the question of why the choice was made, you can never be sure whether there is a real need for privacy or whether some other factor, like shyness or a headache, was responsible for their silence.

Then there is the tricky question of stress. In physical science, it's an easy quantity to define, since it represents the force per unit area applied to a given object, but in mental terms this definition is meaningless. We say that somebody is subjected to stress when he is forced to undergo any experience which makes him mentally uncomfortable. The experience may involve listening to somebody who is saying something he doesn't like, watching an unpleasant scene, or may be concerned with one of the other three human senses.

Alternatively, it may involve thoughts which are unhappy,

frightening, or unpleasant in some other way. The degree of stress will be influenced by the mental state of the person undergoing it and also possibly by his state of health, his tiredness, or similar physiological variables. Thus, any measurement of stress, assuming we can find one, won't be likely to produce consistent results. Typical ways of obtaining a value include exposing the subject to varying levels of stress-causing experiences (such as noise or flashing lights) and noting the point at which some predetermined indicator of irritation sets in. An alternative technique, still in the preliminary stages, is to measure some function of brain activity during the stress application, but, as we still know so little about the brain's functions, it's difficult to see how values obtained in such ways can be interpreted.

The degree of embarrassment is similarly difficult to assess accurately. After all, how do you ask somebody if he's embarrassed? Unless he knows you well (in which case the experiment isn't really valid) he'll be too embarrassed to tell you. In a case of this kind, an experimenter has to rely on secondary symptoms, such as blushing, stammering and movements of the subject's body or hands, to estimate his state of discomfort. As we saw earlier, there are other reasons why people do things like blushing or shuffling, so again the measurement method isn't completely reliable. The situation, too, can influence the degree of discomfort. A person exhibiting all the signs of embarrassment on one occasion may be completely at ease in apparently identical circumstances at a different time if his own mental state has changed for some reason in the interim.

The last of these psychological factors I want to consider is one which leads us nicely into the physiological ones, which is the phenomenon of pain. Basically, it's an unpleasant sensation, caused by overload of the sensory system, which produces symptoms of intense discomfort in the body or mind. It can occur anywhere and is often associated with injury or disease. Its onset may be either gradual or sudden. It may be short-lived or persistent. It may be real or psychosomatic in origin. It's produced by a wide range of stimuli. Originally, scientists assumed it was a result of over-stimulation of any of the other senses, but recent research indicates that it may be associated with a separate cutaneous sense with its own neural system.

Different stimuli, such as a hot wire, an electric current, a tug on a single hair, or the jab of a needle, all give identical sensations to a blindfolded subject. If the stimulus is very short-lived, a "pricking" sensation is reported, while a more prolonged exposure is sensed as a "burning" pain in all cases. There are two distinct types of pain: skin and deep. Skin pain disappears quickly if the stimulus is removed, but deep pain in muscle, joint, bone or organs (e.g. gall stones or kidney stones, appendicitis, ulcer, dislocation, sprain or fracture) is considerably more unpleasant and persistent. It's sensed as a dull ache and is much harder to localise than a skin pain. It can also produce side-effects like a slower heartbeat, lower blood pressure, nausea or sweating. This behaviour should be contrasted with the reflex response to skin pain, a response that causes the stimulated part of the body to move quickly away from the source of pain. Thus, pain is a warning of some threat to the body and is a necessary function for survival. In its absence we'd be in danger of losing health or life without being aware of any need to safeguard ourselves against the hazard to which we were exposed.

Psychologists talk in terms of a pain threshold, which is an impressive way of saying that we don't feel pain until it gets big enough to be brought forcibly to our attention. According to the theories invoking this parameter, each individual has his or her own characteristic pain threshold, or level at which pain is first sensed, and this level can be adjusted to some extent either by mental exertion or by the use of painkillers. The difficulty of measuring such an elusive quantity is, I imagine, fairly obvious. Pain intensity itself is difficult to measure, and different parts of the body have different sensitivities. The "hollows" of the body are more susceptible to pain than the "bumps," and the end of a limb furthest from the body has lower sensitivity to pain than the end nearer to the body (in direct contrast to the sense of touch, it should be noted). Thus, to attempt to quantify a value of pain threshold is virtually impossible. At best, we can only really hope to find a point at which an applied force causes pain to be felt, because the level of pain, as well as being impossible to define in quantitative terms, may change as a result of dulling of the senses by the brain's defence mechanism, even when no change in the pain-causing activity takes place.

It used to be thought that pain extinction shouldn't exist, because such a mechanism (being a warning of risk) would be unfavourable to survival of the species, but modern theories don't agree with this hypothesis. It's found that, if care is taken to fix the level of a stimulus (the needle pressure, source temperature, etc.), then pain eventually subsides to undetectable levels. In normal circumstances, the level isn't maintained constant, as a result of movement of the body, blood circulation, and similar activities, so the continual change in stimulus level ensures that pain sensation doesn't disappear. Pain may also be transferred to other parts of the body than where it actually arises; examples include the pain of angina shooting down the left arm, the tingling that results when your funny bone is struck, and the pain in the part of the body affected when a slipped disc pinches a nerve.

Now to the question of how we measure physiological factors relating to comfort. Let's get away from the subjective techniques and have a change of scenery for a while. The first of the body features which we can attempt to assess is that of temperature. This seems as though it should be a fairly simple task at first sight, but there are dangers in making this assumption. The major one stems from the fact, discussed in the last chapter, that the human thermoregulatory system is prepared to sacrifice odd bits of a body for the sake of survival of the whole. As you'll no doubt recall, one of the ways of doing this is to shut off the blood supply to the less necessary bits, by the process of vasoconstriction, and allowing them to cool down so that heat at the core may be preserved.

Thus, if you happen to measure the temperature of one of the redundant bits at a time when the body is debating the possibility of throwing it to the dogs, you'll get a low reading. It may, in fact, be as much as twenty or thirty Celsius degrees different from the core temperature before permanent damage sets in. What we need to measure, then, depends on what reading we're seeking. Do we want to discover the actual temperature of the organism as a whole? If so, we need to find body core temperature and do so by placing a thermometer in the subject's mouth (if we're sure that little variation exists at all regions of the body) or in the subject's unmentionable other end if we suspect the presence of wide variations.

On the other hand, we may wish to know how temperature varies at different parts of the body as part of an experiment to establish the efficiency of blood flow, for instance. In this situation, as well as a rectal thermometer to check core temperature, we also need to use several temperature-measuring devices (such as thermocouples or thermistors) taped to the external surface of the body at strategic locations. The snags arising from this type of measurement are twofold; the temperature measured is likely to be an intermediate one somewhere between body and environment, and the body temperature may be modified by the presence of the extraneous device (as a result of skin irritation, for example), so readings must be treated with some reserve. Usually, however, modern techniques allow these potential sources of error to be minimised in taking such measurements, and it's now possible to produce a reliable estimate of how temperature varies at different sites. With a core temperature of 37° C, isotherms like those in Figure 22 can be plotted and show that, on a cold day, typical values of temperature are 36° C at the shoulders, 34° C in the thighs, 32° C in the upper arms, 28° C in the calves, 24° C in the lower part of the arms, 20° C at the ankles or wrists and as low as 16° C in the fingers or toes.

The second purely objective measure is that of lung function or efficiency. This is an estimate of how much free volume (for the intake of air) the lungs contain. The value can be changed by such factors as state of health, exposure to contaminants (tobacco smoke is perhaps the most common one) which can reduce the available absorption power, and hereditary or environmental ones. In principle, the estimate is obtained by having the subject draw a deep breath, then exhale it into a calibrated receptacle. In practice, once again, the measurement can be complicated by experimental conditions. The subject may not take in enough air to fill completely the available lung space or may not exhale all that has been taken in.

The rate of exhaling, too, can influence the total volume measured, a rapid rate tending to give somewhat higher amounts. If the subject has a cold, or any similar problem affecting lung function, the apparent capacity can decrease considerably. It should be realised that we can't empty the lungs completely, because a residual amount (the alveolar air) is always retained within their pores. There are, however, three stages of breathing. The tidal air, which is the amount usually exchanged, may be augmented by complemental air in deep breathing and by supplemental air if forced breathing

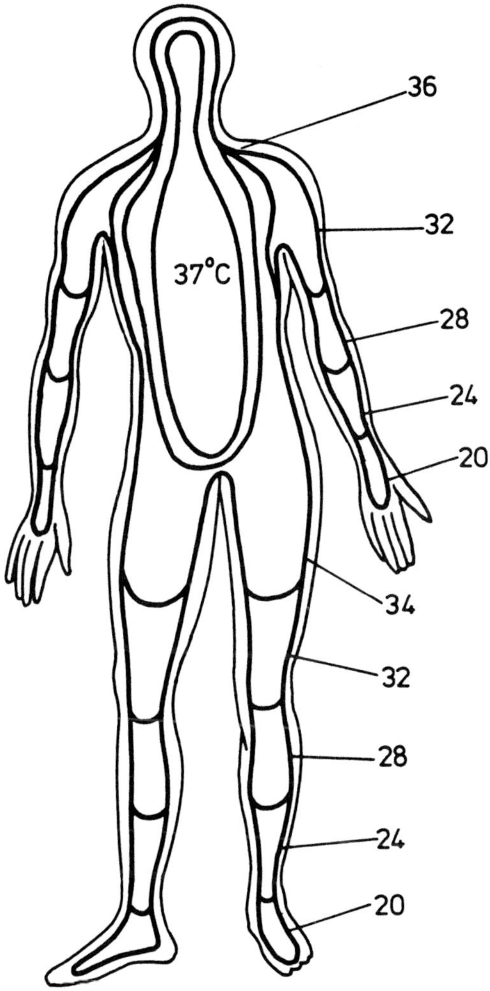

Figure 22. Isotherms on the human body.

is used. The tidal air volume is normally about 500 ml, and the other three quantities (alveolar, complemental and supplemental) are of the order of 1500 ml each, so it's obvious that considerable variation can arise as a result of minor differences in breathing style.

Measurement of blood pressure is also essentially objective. The technique is reasonably familiar to anybody who has ever had a medical examination and involves a process of cutting off the blood supply from a portion of the body (usually an arm) by means of an inflatable cuff. The point at which arterial circulation to the arm beyond the cuff just reappears during heart beats is registered (in millimetres of mercury) on a gauge and is known as the systolic pressure. Then the cuff is allowed to deflate and the point at which blood flows freely again, during both the heartbeats and the resting intervals, is read as the diastolic pressure. In the body, the pressure varies between these two extremes during a normal heartbeat cycle, and the blood pressure reading is always quoted as a ratio between the figures.

In healthy adults, values from 120/80 to 150/90 are typical but may not be constant. Thus, a measurement is again complicated by uncertainty about the validity of the readings obtained. Blood pressure is increased by muscular exertion, emotional excitement, and even by minor changes in such external factors as temperature or altitude. So, if you're being chased up a high mountain on a hot day by a pretty nurse who has been told to take your blood pressure, the value obtained should be regarded as suspect (as also should the mentality of anybody foolish enough to run away from a pretty nurse!)

The last purely objective measure of physiological comfort I want to include is that of body mechanical efficiency. This involves an estimate of the ability of the muscles to do work. In theory, it's no more complicated than any other test to perform, but the measurement is made difficult by the fact that most of us don't possess isolated muscles. In other words, if I'm trying to find out how efficient a particular muscle group is, how can I be sure that no assistance is being provided by other nearby muscles whose effect I want to exclude? This type of measurement has only recently become accurately possible, mainly because of the advent of sophisticated machines which can isolate individual muscles (or muscle groups) for exclusive study. They're expensive and complicated pieces of equipment, requiring expert technical skill, and operate by ensuring that all extraneous muscle aid is rigidly excluded by the clamping arrangements adopted.

Once more, the value obtained in a reading may be influenced by factors outside the knowledge or control of the experimenter.

Muscle efficiency is influenced by the state of health, the muscle tone, or condition of fitness, and the history of use of the particular part of the body. A cyclist, for instance, will have stronger leg muscles than a bricklayer, who stands still at his work, but the bricklayer will tend to have much stronger arm muscles. For an overall efficiency, you must either lay bricks whilst cycling round the wall or find some other way of bodybuilding.

Now for the physiological measurements which combine some degree of subjective and objective testing. In these, a precisely-measurable stimulus is applied to the body and a response (usually gauged by some kind of change or comment on the part of the person) is detected. The most familiar type of test is the visual one, where an optical specialist asks you to tell him what you see in a picture, or if you can read a row of small letters on a card. The first test gauges colour-blindness and the second one tells something about the specific accuracy quality of your sight. Measurements can be influenced by such factors as general fatigue, recent eyestrain, or injury, and the importance of the sight sense makes it imperative that any defects are identified for correction as soon as possible. The tests depend to some extent on the honesty of replies given by the subject, and it's clearly pointless to expect reasonable results without total cooperation.

Hearing tests are similar in nature. The ears are covered by a headset containing earphones and sounds are reproduced at the subject's ear. The lowest audible amplitude of a sound note which can be detected is recorded and a chart constructed showing the variation of threshold of audibility with sound frequency as in the example shown as Figure 12, in Chapter 5. This chart is then compared with one showing the expected performance for a person of the same age, as in Figure 13, and any deficiencies are noted. The measurement, as is implied by the procedure used, is affected by age. Most of us experience some loss of hearing ability, particularly at high frequencies, as we get older. It's also influenced, however, by the environmental sound level to which the person has been subjected. Long exposure to loud noises can reduce hearing ability considerably, as is evident if you've had any dealings with workers engaged, for example, in the weaving of textile fabrics. Deafness may increase imperceptibly over a period of years if the noise is consistently at a medically unsafe, but not dangerously high level, or hearing may be lost instantly if an extra loud noise, such as an explosion, is encountered.

Tactile sensitivity and neural response are a little more difficult to measure. For these two, you're trying to isolate the response of one nerve (on the skin or within the body respectively) to a given stimulus, and the body is notorious for playing tricks on experimenters in such esoteric conditions. In theory, you apply a stimulus and measure the response obtained, either by asking the subject to tell you when he feels anything or by checking some kind of reflex action made by his body. In practice, it's not usually that simple. For a start, the subject may tell you, unwittingly or otherwise, that he feels a sensation when in reality he's only imagining it. The sense of touch can easily be deceived, and the presence of a stimulus may not be noticed. If you decide to go for an objective measure to eliminate the risk of spurious subjective results, how do you isolate the response of a single nerve?

One of the techniques to overcome this kind of difficulty is to use the brain as a test region, by inserting electrodes or with external detectors, as in an encephelograph. These techniques, though, are still in the early stages of research, are not yet completely understood, and are technically cumbersome yet delicate to use. Experiments on animals, in which single nerve cells are isolated surgically and the electrical response to various types of stimuli recorded, can provide an appreciable amount of information, but the validity of results when transferred to human beings hasn't been established convincingly to date. Much work remains to be done in this area on the frontiers of science and will no doubt reveal further secrets about the operation of our bodies in the years ahead.

The senses of taste and smell are even more difficult to investigate experimentally, since it's difficult to apply a fixed amount of stimulus when we're not sure how that stimulus itself is measured. How, for example, do you know that the taste sensations of two different people will always bear a constant relation to one another when you can't be sure that a given individual's perceptions remain constant? Try drinking a medium claret before and after your dessert if you want to discover how your taste buds can lie to you. With the lingering taste of a savoury course in your mouth, the claret may well taste quite sweet, but after the sweetness of dessert has coloured your palate, the wine will appear almost bitter.

This type of problem is common to all sensory testing. It's impossible to know whether two people see, or hear, or taste, or smell, or feel the same sensation under identical conditions. All that can be done is to ask the subjects to perform a comparison of stimuli. There are three techniques in common use for this purpose, these being discrimination, indirect scaling and direct scaling. In the first of these, the subject is presented with two stimuli and asked to say whether or not they're different. Alternatively, he may be asked to make changes to one stimulus until the two are identical, or asked to state whether a variable stimulus is present or absent, given a standard one for comparative purposes.

In indirect scaling, several pairs of stimuli are compared and a decision made as to which is the larger in each case, after which all the stimuli are arranged in order (possibly with some ambiguities!). Alternatively, each stimulus may be rated in isolation, either by a simple rank-ordering or by allocating a number, on a scale defined by the experimenter, to each one. In direct scaling, a magnitude estimation is made with reference to a standard value. If, for instance, the standard is assigned a value of ten and the unknown is perceived as twice as large as the standard, it would be assigned a value of twenty. As before, alternative techniques are available and include asking the subject to adjust an unknown stimulus to have, say, half or twice the value of the standard one. An interesting variation on this is to ask the subject to match one stimulus by means of another; if, for example, he's presented with a sound, he must then match it with a sound of a different frequency or with a light. For obvious reasons, a detailed description of all these tests isn't within the scope of this book, and the reader is referred to the reference works listed at the end of the volume for further details.

Thus, tests of sensory evaluation are just beginning to be developed. For taste and smell, the science is even less advanced than for the other senses. The investigator must usually rely on a comparison between the unknown taste or odour and established ones which have some degree of standardisation associated with them after long trials. So, for instance, you ask your panelist to rate the unknown taste as "sweeter than" or "more acidic than" a standard and then judge the contrast between a range of unknowns by obtaining some kind of ranking order from all the results. The process is a tedious, frustrating one, and the values obtained mustn't be analysed quantitatively, since we've no way of knowing whether a given number assigned by one individual has the same meaning as an identical number assigned by the next person in the test panel. Once more, we're faced with the difficult task of attempting to carry out a measurement with no knowledge of how the "instrument" (i.e. the human being) is functioning during the measurement. One recent development which may be invaluable in this area is a measurement of the electrical activity resulting within the brain when a stimulus is presented. This is currently feasible for stimulation by light and may in the future be extended to enable the other senses to be monitored.

Finally, the question of skin hygiene belongs somewhere in this chapter, and this is as good a place as any to put it. Skins are sensitive to a variety of potentially troublesome ailments of a very minor kind which can affect comfort out of all proportion to their magnitude. Examples of the sort of thing I mean include odour, acne, dryness, itch, and all the other problems of similar kind which afflict the human race. (Itch may be defined as the sensation felt when stimulation of pain receptors is present in too small a magnitude to produce pain.) These problems arise because the cells of the skin experience some kind of adverse reaction when they come into contact with the source of trouble.

Perhaps bacterial infection should be mentioned first, since it's often the root cause of other problems. Bacterial growth is encouraged by conditions favourable to the cell division necessary for culture to occur. Body secretions can influence the rate at which growth takes place, as also can the metabolic rate of individuals, by changing the microclimate at the surface of the skin. The task of measuring these minute differences, however, is immense and usually not even considered. Instead, we either try to simulate the conditions prevailing at the skin or carry out an estimation of the number of bacteria present there after a given test period. This is done by collecting test specimens and allowing the bacteria in them to grow under carefully controlled conditions until they become visible and can be counted.

Other types of skin sensitivity, usually the result of some allergic reaction to the presence of an irritant or the absence of sufficient moisture, aren't normally measured quantitatively. If they're present to a degree which causes the subject undue mental or physical misery, treatment to reduce pain or itch, or to restore the skin to an acceptable appearance, is carried out, but otherwise the problem isn't addressed until we're forced to do so. As before, the difficulties of measurement are immense; there is no realistic way to check the quantitative magnitude of an allergic reaction. Either it's bad or it isn't. If it is, we try to do something about it. If it isn't, we tolerate it.

Now, we come to the problem of measuring the physical properties relating to comfort, and a precautionary word is in order at this point. So far, all the measurements we've carried out have a direct bearing on the comfort of the body. Even when the techniques were objective, we were assessing what was happening to the human being when we took a reading. In the remainder of this chapter, an important difference exists. The laboratory testing of physical properties is carried out on some material or object which isn't a part of the human body but is indirectly associated with it when comfort matters are important. In other words, we measure, for example, the ability of a substance to retain or transmit heat, then predict how good it will be when used as an insulating material. Notice that we can't be sure that our predictions will be correct; that can only be done by actual tests on the body when the article is used for insulation.

This is part of a general phenomenon accompanying all simulation testing. We set up a laboratory experiment designed to tell us how a given article will perform in a practical situation and make assumptions from the results of our tests. We must then carry out a field trial of the article, under normal conditions of use, to discover how good our predictions were. For most laboratory tests of this kind, there is usually a relatively poor correlation with practical trials, and results can be misleading. For this reason, it's customary to make no claims about the quantitative properties of an article but only to attempt a rank-ordering. That is, we don't say how good two different articles will be in a particular application but only which of the two is likely to be the better.

Why, then, do we carry out this type of testing if it's so inaccurate and inconclusive? The answer is simple. The test may not be accurate, but it does give us some guidance with which, in the light of experience, we can make fairly confident predictions of how a particular article will behave under specified conditions of use. The predictions may be wrong on occasion but are usually reasonably close to the mark. Secondly, the results of tests of this type are much more reliable, and obtained much more cheaply and easily, than are tests involving human subjects. Because the tests are objective ones, fairly high precision of measurement is normally achieved so that we can be more confident about the true value of a result. Speed of testing, which may take years with subjective trials, means that we can often get results within days or even hours. If the testing is carried out by one person (as is usually the case), there is only one wage to pay, and it's not necessary to hire people as subjects for extended periods of time, as happens with field trials. Thus, with all their drawbacks, objective laboratory tests are normally greatly preferred to subjective ones, and we should therefore consider them very seriously in attempting to assess comfort properties.

The physical factor of greatest importance is probably the ability to allow or prevent the flow of heat through a barrier. This property is fundamental in establishing the protective ability of a garment intended to keep the body warm (and hence alive) in cold weather or cool in hot climates. As we shall see later, the thermal resistance of a garment isn't the only factor which governs the body's ability to retain heat, but it's certainly one of the major ones and may be the most important of all. As is often the case, the measurement of thermal resistance is complicated by several factors. In theory, we establish zones of high and low temperature, then monitor the flow of heat between them. The "unknown" barrier is interposed between the two zones, and the effect on the rate of flow of heat is then assessed. Figure 23 illustrates the principle of operation of the standard method of measurement used for assessing thermal resistance.

A hot region is created in the upper plate by circulating a heated fluid (usually water or oil) through the hollow interior, and the cool region at the bottom is merely a block of metal at room temperature. Immediately next to the hot plate is a layer of cork, or other insulating substance, to make sure that the heat is distributed uniformly over the entire area of the test specimen. A heat meter, a device to measure the rate of flow of heat, is placed immediately next to the

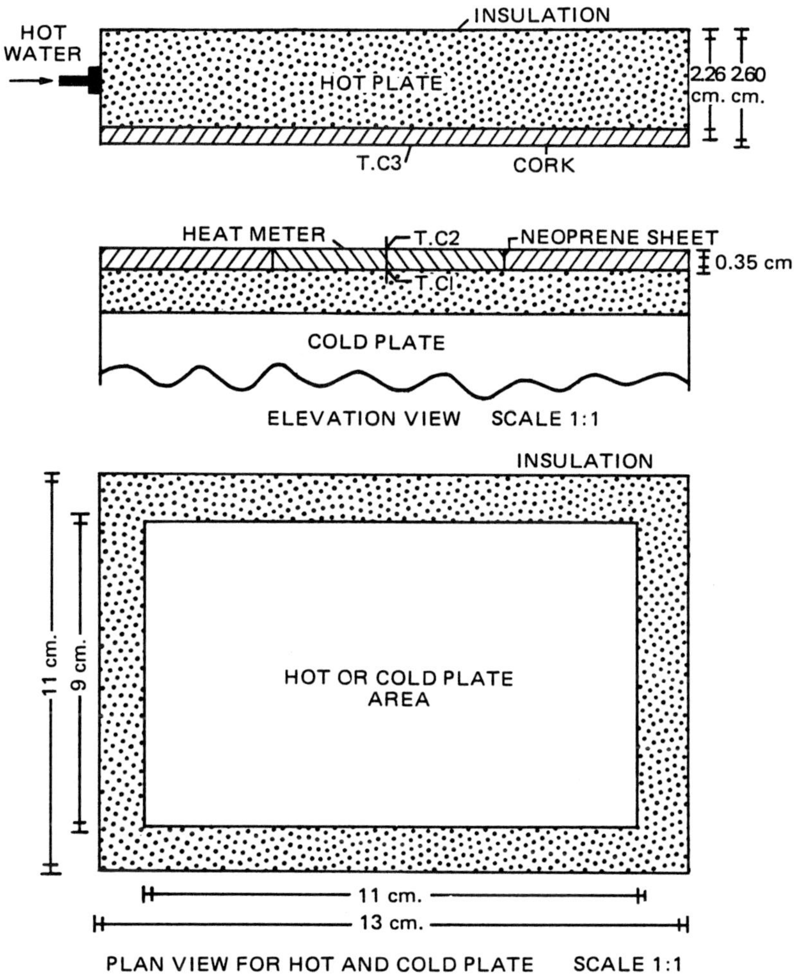

Figure 23. Measurement of thermal resistance.

test "unknown," and several thermocouples are inserted in the heat flow path to monitor temperature at each point in this path. The principle of operation is a very simple one. We know the rate of movement of heat through the system, from readings taken on the heat flow meter, and we know the temperature difference across this meter and across the unknown barrier. It's thus an easy matter to calculate the rate of flow of heat through the barrier and, by measuring its thickness, to calculate its resistance to heat flow.

The problems accompanying the measurement require careful experimentation to ensure their elimination as far as possible. The first of them is the risk of heat flow via routes other than the direct one through the barrier. Heat can escape from the system by leaking from the sides, for example, and this gives an error in the estimation of how much flows along the path you're monitoring. The usual way to overcome this drawback is to include a "guard" arrangement (an additional area of plate surrounding the one used in measurement, so that heat leakage takes place only from this unmonitored region) and to provide large amounts of insulation around the entire apparatus so that any heat which does escape from the sides can't travel very far.

Another more important problem is the fact that the plates are arranged in a horizontal, layered manner in the original apparatus. For most purposes, this isn't a critical defect, but for the measurement of comfort-related properties it creates errors. Suppose, for example, we need to know the thermal insulating ability of an interlining material for winter clothing. As we've already seen, it's the enclosed air spaces which give a garment its insulating ability, and these air spaces are enhanced by a thick, porous structure. Unfortunately for our experiment, such a structure will also be very compressible, so that it will be flattened by the weight of the plate used in the thermal insulation measurement. Thus, part or all of the air responsible for the insulating ability will be squeezed out of the material, with the result that it will appear to be a far poorer insulator than it actually is.

One way in which this problem has been overcome is to turn the apparatus on its side, as shown in Figure 24, where the latest practical version of the equipment is shown. I'm rather biased about the value of this piece of apparatus, since I designed it, so I hope you'll forgive me if I appear to be less critical of its quality than you might wish. As far as I'm aware, though, it suffers from no serious defects which render readings of thermal resistance suspect for any type of material. No doubt I'll be proved wrong in due course, when some interfering busybody of a rival performs critical testing on an outlandish material which no one would ever be insane enough to use anyway. The one slight difficulty in obtaining a measurement, which I include just to show you what an honest, unbiased scientist I really am, is the need to set the fabric thickness on the gauge before measuring a value of heat flow. If the setting is too small, compression occurs and you're back to the same problem as with the original arrangement. If the setting is too wide, heat can escape via the sides of the test specimen and errors are introduced. Thus, a careful measurement of thickness has to be carried out initially.

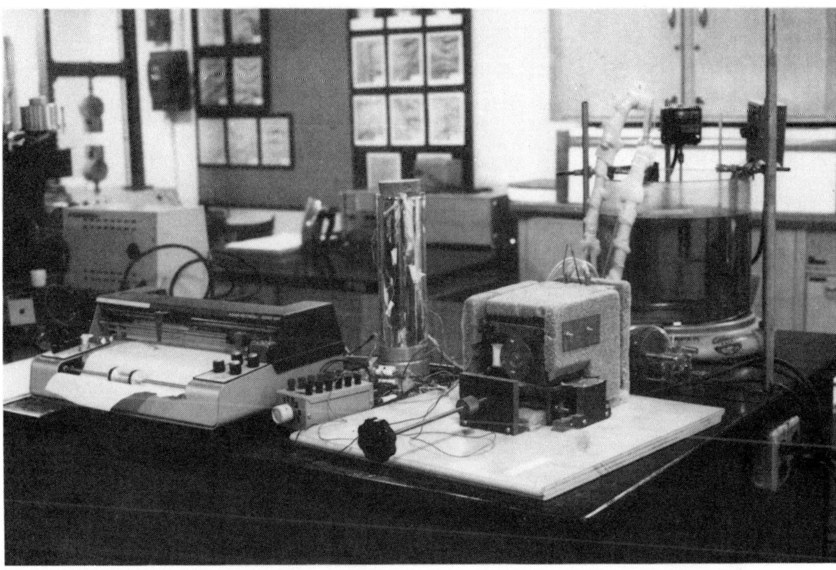

Figure 24. Modified method of measuring thermal resistance.

Thickness is also important in its own right as a comfort-related factor, so we may as well examine its method of measurement in some detail. For a solid material, like a sheet of wood or metal, the procedure is simple. You merely get a ruler, find out how far one face is from the other, and write down the answer. Difficulties can arise if the thickness isn't uniform, but you can usually get an average value that means something. In comfort applications, though, the problem of compressibility, as we've just seen, complicates the situation. Most comfort-related thickness measurements will be associated with fabrics, which can be squashed to a flatter state, so we can't just put a plate on each side and measure the distance between the two plates as we would for a solid.

Instead, we adopt an extrapolative technique. We deliberately apply a load to the fabric, using apparatus like that shown in Figure 25, and measure the distance between the two plates of the tester at that load. Then we apply a series of different loads and repeat the process several times. Finally, we plot a graph showing how thickness varies with load and estimate from the graph what the thickness would be if the load were zero. This is the so-called intrinsic thickness, which we use to represent the true thickness of the material and, incidentally, assume for setting the thickness on the thermal resistance tester described above.

Well, that's enough about thickness and heat transfer. Now we

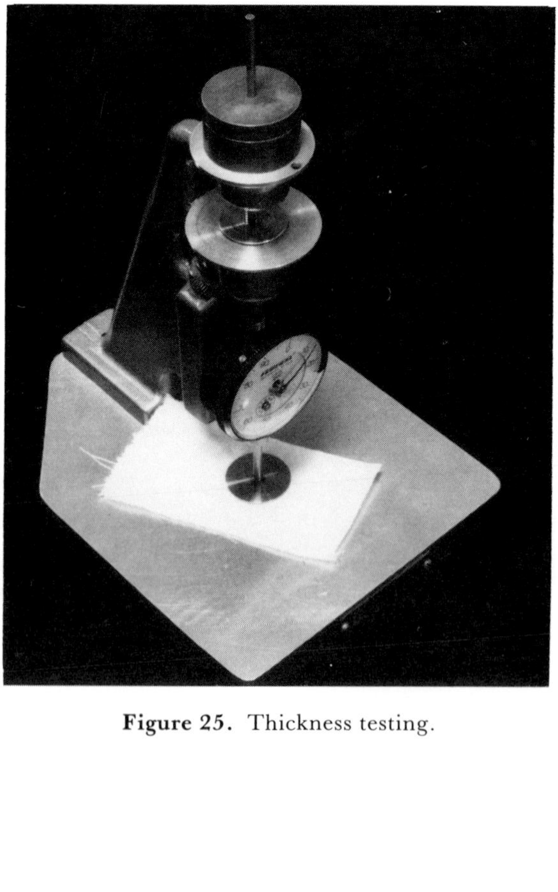

Figure 25. Thickness testing.

turn our attention to a property almost as important — that of moisture movement. There are two measurements to consider here, depending on whether the moisture is in the vapour or the liquid state. Let's look first at the former case, since perspiration comfort depends so heavily on it and is necessary all the time. The measurement of moisture vapour transmission is possibly one of the most difficult experiments to carry out in the whole realm of comfort work, since it's impossible to isolate what you're measuring. Moisture vapour is invariably accompanied by air, and, since the two gases have approximately the same molecular size, there is bound to be some interference in any situation which can easily be envisaged. The theoretical way to overcome this difficulty is to eliminate, as far as possible, the influence of the air.

The traditional technique for performing the measurement is illustrated in Figure 26 and uses a number of dishes (normally six) placed on a turntable arrangement. Three of them are covered with a specimen of a standard fabric only, and the other three have a specimen of the test fabric in addition to the standard one. A precisely

Figure 26. Moisture vapour transmission by the dish technique.

measured amount of water is placed on each of the containers; then the turntable is slowly rotated to avoid any risk of a saturated layer of air accumulating near the upper surface of the specimen and interfering with free evaporation of water. This evaporation is allowed to continue for some time (fourteen hours in one standard test), after which the change in water content within each container is obtained by weighing.

The moisture vapour resistance of the test specimen is then estimated from the difference between the transmission rates of the dishes with and without the test specimens incorporated. Correction factors have to be applied for the change in height and for the free space existing between the water surface and the test specimen, a procedure which can cause significant errors to be introduced into the reading. In addition, the fact that small changes in a large quantity are being measured makes the possibility of errors even more likely. Thus, apart from being time consuming, the method isn't accurate unless great care is observed in taking readings. One of my research students, after considerable practice, was able to obtain reasonably precise results by a full day of hard work on each specimen. He was, however, dubious of the ability of the average technician to be able to develop such skills, in addition to a range of other duties, or to have the time to devote to the careful measurements necessary.

For these reasons, the method isn't a particularly good one, and, in an attempt to find something better, the new technique illustrated in Figure 27 was developed. I trust you'll find the design concept brilliant, since it's again one of mine, but let me describe it for you before you make a snap judgment on such flimsy evidence. Basically, the method involves the establishment of a moisture vapour gradient along a closed tube, by having a reservoir of water at one end and a drying agent at the other. When the system is in equilibrium, water vapour flows from the "wet" to the "dry" end of the apparatus, and the relative humidity decreases linearly along the tube. This gradient can be measured by placing sensors, as shown, at various points in the tube, and the attainment of equilibrium is detected when the readings of these sensors all become constant.

Then, the barrier to water flow is placed in the path and a change in the equilibrium conditions occurs. The gradient becomes

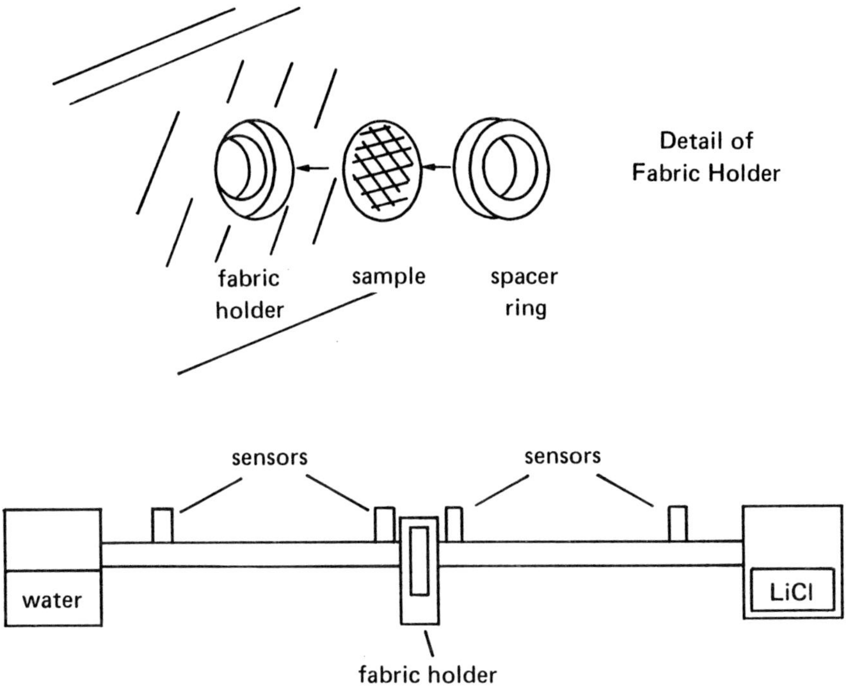

Figure 27. Moisture vapour transmission by the gradient method. (Reproduced by kind permission of The Textile Institute.)

discontinuous, in a manner determined by the impedance to moisture vapour flow caused by the presence of the barrier. When the new equilibrium is established, readings of the change in gradient can be taken and the two gradients plotted together on a single graph, as in Figure 28. The presence of the barrier is equivalent to the addition of a column of air to the system, since the same resistance to flow would be produced by such a column, so the moisture vapour transmission resistance of the barrier can be expressed as the length of this column and is equal to the horizontal distance between the two halves of the discontinuous line. The entire measurement takes up to two hours, depending on how resistant the barrier is, and is at least comparable in accuracy and precision to the traditional one. An enclosed specimen holder is necessary to ensure that no time is lost unnecessarily in waiting for the system to return to equilibrium, but the apparatus is otherwise very simple. It also works well, I'm happy to report, for layer structures (as are usually encountered in comfort work on clothing systems), which the traditional method is totally unable to handle.

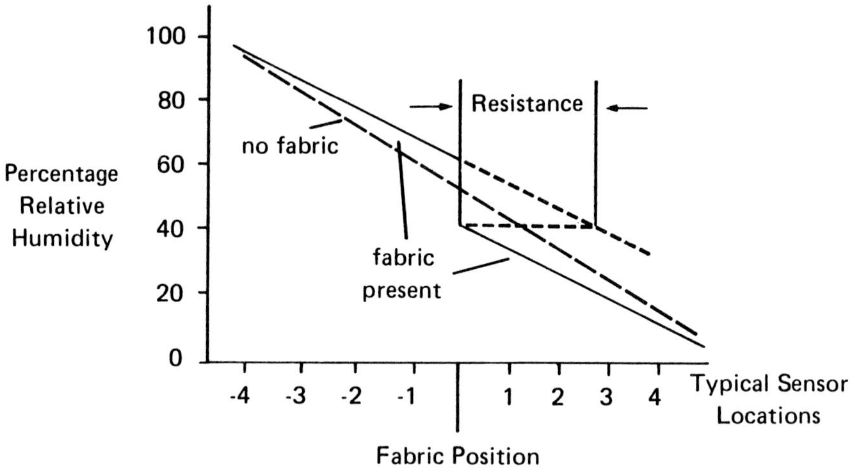

Figure 28. Typical graph obtained using the gradient method of measuring moisture vapour transmission. (Reproduced by kind permission of The Textile Instutite.)

Now for the question of liquid water transmission, which is important in protecting against the discomfort of rain. The idea of the various test methods is to simulate what happens in a storm, so the simplest technique uses a perforated funnel, as shown in Figure 29, to deliver a fixed amount of water, in droplet form, to the specimen under test. There are two ways of assessing the effectiveness of water resistance: a visual and a gravimetric one. In the former, the front and back of the specimen are examined, after a test, to compare their appearance with a photographic standard. A rating of 0 to 100 is given, to represent varying degrees of penetration, with 0 being total saturation of both surfaces and 100 being no wetting or sticking of water to either surface. Thus, a rating of 100 for a fabric would

CGSB
STANDARD
4-GP-2

METHOD 26

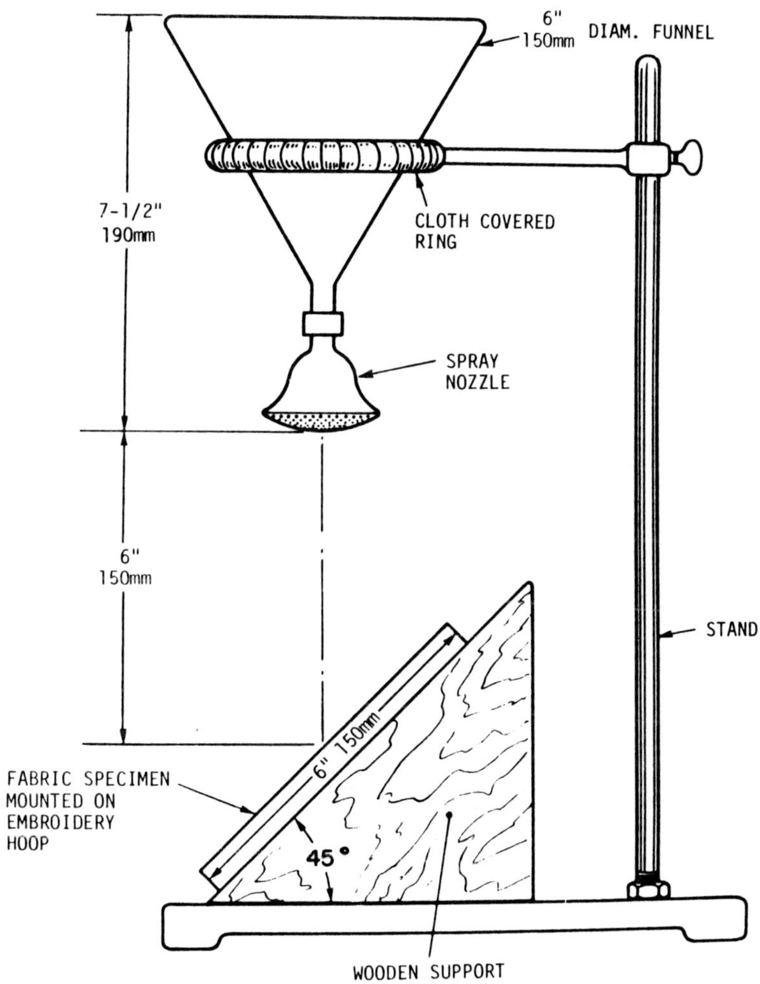

Figure 29. Liquid water resistance measurement by the funnel method (Reproduced by kind permission of the Canadian General Standards Board from CAN2-4.2-M77 — Textile Test Methods, Method 26.2. Water Resistance — Surface Wetting Test (Spray Test).

imply excellent potential rain protection and one of 0 idiocy in the wearer if he ventured outdoors, in a garment made from it, with any hint of cloud on the horizon.

In the gravimetric method, the water dropping on to the fabric is collected, after impact, in one of two possible places. It will either pass through the specimen and be collected in a container placed under the frame for the purpose, or run off and be similarly collected in a second container, or remain within the specimen and not be collected at all. By weighing the two containers before and after a test, the amounts of water which penetrate, get absorbed or are transmitted by the specimen can be compared and conclusions drawn about the fabric's effectiveness in resisting water flow. The test, of course, is only valid under the conditions pertaining. If the rainfall happens not to be of 100 ml from a height of 50 cm (the normal conditions of test), the results may be entirely different. Nevertheless, the test is a useful one, as usual, for comparing the likely behaviour of materials in rainy weather.

Another test worthy of mention uses a charge of water under fixed pressure, in contact with the specimen, as the potential wetting source as shown in Figure 30. The rate of transfer of water across the barrier of the specimen, under the hydrostatic pressure prevailing, is a measure of the resistance to liquid water flow, but the test isn't really appropriate to the conditions expected in rain. It's more useful for predicting the likelihood of water seeping through, say, a diving suit fabric, a situation which may well be important for the continuing comfort, not to mention survival, of the diver dressed in it.

The final property I intend to mention in this chapter is that of wind chill, which I've left until last because it's difficult to see how it can be measured effectively. Some work in the area is taking place sporadically, but no one (not even me!) has yet devised a convincing way to estimate its magnitude.

There are two approaches to the problem. The first of these measures air permeability, using apparatus of the kind shown in Figure 31. A fixed pressure is established between opposite faces of the fabric, and the rate at which air flows through it is assessed by the instrument. The reading, of course, is virtually independent of temperature, so it gives no real guidance about the fabric's behaviour in actual wind chill conditions, and an alternative method of assessment is needed.

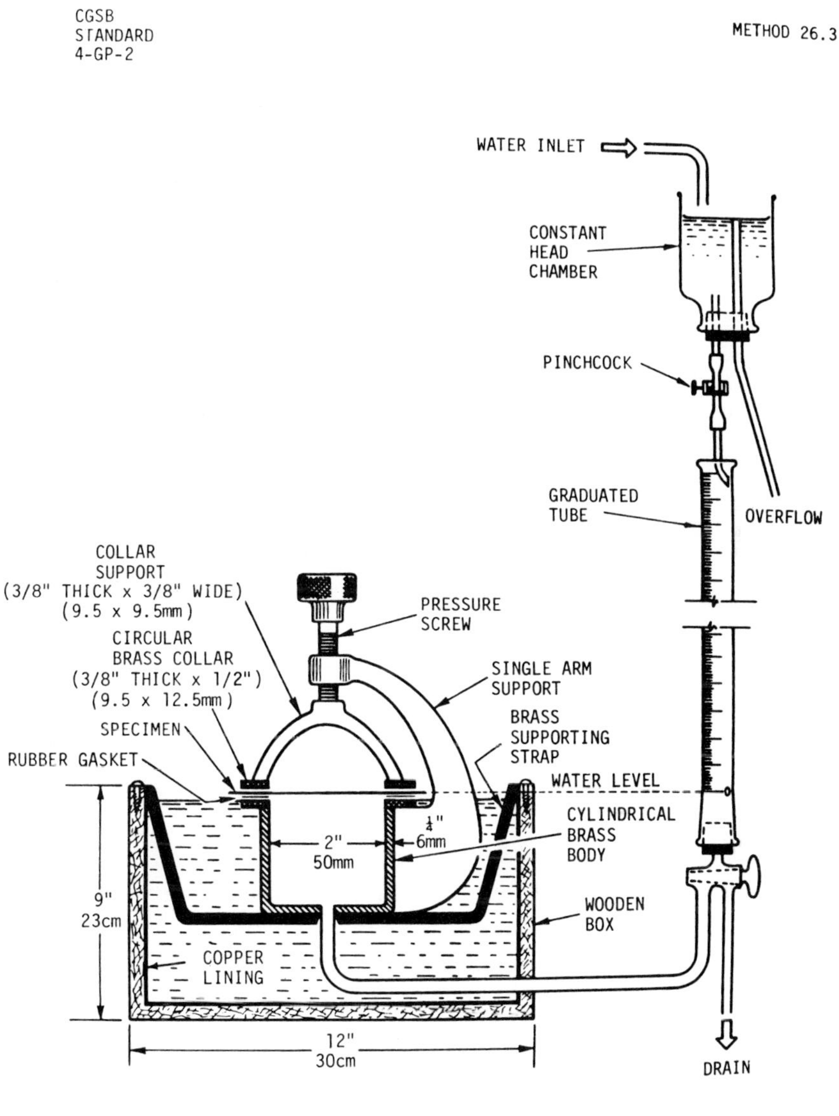

Figure 30. Liquid water resistance measurement by the hydrostatic method. (Reproduced by kind permission of the Canadian General Standards Board from CAN2-4.2-M77 Textile Test Methods, Method 26.3 Water Resistance, Variable Head Penetration Test).

Figure 31. Instrument for the measurement of air permeability.

What is normally done is to measure the temperature and wind speed, then compute a theoretical figure which is supposed to represent the wind chill factor. This is a number, recorded as degrees Celsius, giving the temperature drop below the prevailing one which would make you feel just as cold in still air. It's obvious, even from the bald statement of this definition, that the measure is full of suspicious fudge factors. An equation often used by the United States Army, government agencies, and others involved in cold-climate work has been developed as shown below and relates the heat loss in kilocalories per square metre with the wind speed v in metres per second and the air temperature t in degrees Celsius.

$$\text{Wind chill} = (10.45 + 10 \times \sqrt{v} - v)(33 - t)$$

Clearly, the actual effect of wind chill will depend on many characteristics which are individual in nature, such as the physiology and low-temperature tolerance of the person who's supposed to be feeling the increased cold. The extent of the effect is influenced by the relative movement of the air around the human being. A garment which has excellent insulation properties in still air, for example, can be useless in windy conditions, since the "dead air" layer it contains may then be swept away if the structure is an open one. Athletes running in winter are particularly prone to wind chill for this reason, especially if they're sweating and thus losing heat by reduction of available insulating air space as well. It should therefore be obvious that the measurement of wind chill will also be influenced by the time and extent of exposure, the clothing worn, and even the direction in which you're facing, so the quantity being measured is a vague one at best. Add to that the fact that scientists can't agree on the best formula to use in calculating the fudge factor, and you have the beginnings of a measurement method which can be fought over by experts for years to come!

These, then, are some of the techniques available for determining the more common properties used in assessing comfort. The process isn't a very satisfactory one, as you'll have gathered, but it's the best one that is available. In the remainder of this book, we'll come across examples, from time to time, of ways in which measurements are used and of problems arising as a result of their use. Let us, then, examine how comfort affects real life!

CHAPTER 7

THE ENVIRONMENT AND COMFORT

IN THE past few chapters, we've often referred in passing to the environment and, indeed, have actually looked briefly in the first chapter at some of the ways in which it can influence our comfort. Because this dependence is so important, I'd like to spend a little time in considering the connections between the two subjects in more depth. In the term "the environment," we normally include everything that is outside the human body and that can have some influence, no matter how slight, on it. But the parts of the environment which can influence comfort appreciably are, of course, much more limited. We can divide our discussion into several areas, beginning with a general one and then considering in turn such factors as the effect on the senses, aspects of air, water and land pollution, and the way in which mankind has affected his surroundings to make them more (or less!) comfortable for himself.

We should first examine a nebulous subject which is usually expressed as the "quality of life." By this term, we normally mean those aspects of life which aren't essential for survival but which make living more endurable, and hence more comfortable, in some way or other. Thus, we must have food and water to survive, but the extension of those necessities to steak and a good burgundy would be classed as enhancing the quality of life. Similarly, we need to arrange for breeding to occur if the human race as a whole is to survive, but pair bonding for the sake of companionship makes that survival (in most cases, at least) a more pleasant experience. We need to see where we're going and be able to hear the approach of danger to exist in a modern world, but the opportunity to watch a sunset, or the stars, or a beautiful scene (or a beautiful girl!) makes the value of vision doubly important, and the opportunity to listen to a Beethoven symphony (or a rock group if you really insist) or words of love enhances the hearing sense also.

The fact that the pleasant things of life which I've just described are a mixture of natural and manmade ones is no accident. To be comfortable in our minds, we need to enjoy both aspects of the two types of factor making up this nebulous property called quality of life. In our modern world, we've become accustomed to having the best of both sides of the bargain. We ask for, and usually get, the modern conveniences which give us the time to enjoy more leisure and the natural beauty which we can, as a result, find time to appreciate. No one can be truly comfortable living in an exclusively manmade world; you've only to look at the statistics concerning crime in a big city to realise that all isn't sweetness and light for some members of society.

In the same way, we couldn't be comfortable if we had only the natural world without any of the modern developments to depend on. Picture yourself in a situation where heating, electrical power, tap water, transportation, clothing, groceries, and all the other conveniences of modern life were suddenly removed at a stroke. It's all very well to dream of an idyllic life on some desert island, but a diet of berries, worms, grubs and all the other nasty things our distant ancestors used as food for survival wouldn't be acceptable to most of us. Think, too, of the loneliness, not to mention the painful cuts or scratches on sensitive bits of your unclothed anatomy. Paradise would become purgatory, and you'd soon find yourself longing for the crass delights of that appalling and dreadful society which you'd shunned as too unbearably fast.

One fact which must never be allowed to escape our notice, though, is the need for a balance to be struck in nature as well as in man. If we begin to demand too many of the manmade comforts, then it follows that we must accept a diminution in the natural ones. In the first place, there won't be time to spend in the pursuit of innocent pastimes in the open air if the world is full of alternative pleasures. Even if we're weird, and make the time by refusing to have anything to do with these new-fangled fads, it may well become impossible to find natural beauty any more when the world is bent on the pursuit of manmade comforts. Also, and much more important, there's the fact that we must rob nature in other ways to pay for our novelties. We extract the resources from the ground and squander them wholesale. We push nature aside to make more room for the roads, the houses, the shops, the factories, and all the other vital support systems which can't be omitted from our modern life-style.

The balance of nature, that most precarious balance which has been so carefully achieved throughout millions of years, stands on the brink of collapse. In creating the fool's paradise that we call civilisation, we're gradually destroying the world around us, so that it will one day be unfit to support our race. Scientists persist in sounding the warnings of doom, but they do so after driving their car to the hall where the pronouncement is made and then go home to their big, air-conditioned home in the smog-filled air of an ugly city to watch television programmes giving further dire warnings of the erosion of the earth's resources. Don't think that I excuse myself; I sit at my computer keyboard typing this manuscript on paper which has been stolen from nature and has been produced by burning irreplaceable fuel, not to mention the chemicals consumed in the manufacturing process. Nature isn't infinite. Every bit of "progress" we make has to be paid for in some way, and the cost is increasing at an ever more rapid pace. Yet, many of us preach the message of despair without being able to stop our own small share in hastening mankind's headlong dash towards extinction.

Perhaps the most fashionable of the many ways in which mankind can exert an influence on the natural world around him is the phenomenon normally summarised in the word "pollution." The newspapers are filled with stories of how man's polluting his planet, and the number of consumer groups advocating some form of pollution control (in North America at least), is legion. There are many forms of pollution, and I want to consider in detail, in this chapter, five of the most pervasive ones. The first one on my list isn't usually thought of as pollution by most people, but it happens to be one which I find infuriating and, since it's my book, I have the privilege of choosing how I order my material.

The particular form of pollution which I detest is the visual one. By that, I mean the deliberate act of inflicting on the environment material which is ugly in the eye of the beholder. I could list many examples of visual pollution which bother me, but I'll restrict myself to a few of the worst ones and leave you to find others which arouse similar feelings of discomfort in your own mind. I loathe the sight of waste material in the form of plastic, paper, cartons, bottles, rusty metal, and all the other miscellaneous cast-off pieces of civilisation which appear wherever people gather or move in the mass. Our parklands, our roads, our forests or fields, our trains or buses, our sports arenas, our shores or mountains, even our gardens, are littered with rubbish carelessly thrown away by unthinking morons who have no sense of mental tidiness and no concern for the natural beauty in which this world abounds. I know the material isn't physically harmful to me (unless it happens to contain rotting food which can cause disease or attract various animals) and that I'm almost a lone voice of protest in Canadian society, to judge by the number of offenders I see, but that doesn't make me feel any less uncomfortable when I see garbage strewn around the world.

If I could persuade enough people to share my viewpoint, I'd make the act of littering the public domain with "harmless" material an offence punishable by severe penalties and with no leniency for first offenders. I venture to suggest that such penalties would not only clean the world but would also make it a better place to inhabit in other ways, also. The act of creating visual pollution is symptomatic of a total disregard for the feelings of others, and if parents were made to pay for their children's neglect of a simple act of courtesy, they would perhaps instill into their offspring a sense of altruism which would manifest itself in so many other unexpected ways.

How can you expect children to have pride in their city, though, when their elders fill its streets with garish signs, wires, traffic, scrap vehicles, the cast-off detritus of their so-called civilisation? And how can you expect the population at large to become educated to the dangers of this continuing attack on nature when their political leaders are prepared to turn a blind eye to just about any form of pollution as long as it brings in money or votes? Truly, we live in a world where sanity is a scarce commodity!

Well, having aired my own pet grievance, let me get on to the more popular ones, the ones which do cause concern in other people besides myself. The first of these is air pollution, the act of filling the atmosphere around us with noxious, toxic, or otherwise undesirable substances. Even in a healthy atmosphere, one which would be perfectly acceptable by any standards, discomfort can arise. Heat can produce a mirage, or perspiration. Cold can give us sore eyes and make us shiver. Dryness produces thirst and dehydration. Humidity causes perspiration or psychological discomfort that can lead to the kind of riot seen occasionally in some of the large cities of North America in the heat of a humid summer. Small wonder, then, that we're easily made uncomfortable when the environmental conditions fall short of being satisfactory, usually by man's direct interference.

The current newsworthy form of air pollution, at the time of writing this chapter, happens to be the acid gases, emitted by various industries, that cause acid rain. I don't suppose most of the general public gave a thought to the problem until it began to affect the relations between Canada and the United States of America, but it's been around for some time, ever since mankind began to make fire, in fact, though it didn't create much of a problem back in those good old days. A significant jump in the world's acid gas production accompanied the industrial revolution, when fossil fuels (mainly coal) began to be consumed at an enormously increased rate to satisfy the demand for mass-produced goods in a mechanised society. With the advent of modern affluence, that demand has increased considerably and, as a result, so has the production of the pollutants responsible for acid rain. The same gases, though, have long caused the famous London pea-souper fogs, the choking clouds that coat the inside of British lungs and give the inhabitants of those parts their characteristic national chest disorders.

Acid gases, along with other undesirable compounds, are also released into the air as an end product of the combustion of fuel in the various types of internal-combustion engine used so commonly in our modern world. Automobile, aircraft, and other forms of transportation all need the assistance of engines driven by petroleum byproducts (or are dependent on such fuels completely) in order to operate. Any contemporary machine using such fuels is inefficient; that is, it can't convert all the energy locked up in the fuel completely to usable power. The proportion of fuel converted to power varies from something like 20 percent to 40 percent, depending on the type of engine used and the state of maintenance in which it's kept. The wasted fuel appears as by-products at the exhaust and is emitted into the air either unused or in the form of the pollutants we're discussing.

Naturally, as mankind becomes more technologically advanced, the human race wants to exercise its curiosity by spending more of its available leisure time in seeing more and more of the world in which we live. So, as the need for transportation increases, in direct consequences, so does the waste fuel spewed into the atmosphere. You've only to examine the horizon as your jet aircraft climbs away from the airport to see visible symptoms of the normally hidden atmospheric pollution in the haze hanging over any of the world's large cities. Or, if you happen not to like flying, climb a nearby mountain and look down on the same city to see the clouds of smog, unnoticeable at ground level, which are slowly gathering like a time bomb to choke the inhabitants in a few years' time.

Other forms of atmospheric pollution which have caught the imagination of writers in the past include such substances as the gases used as propellants in pressurised containers or those given off from solvents used in industrial processes. Clouds of nitrous oxide pour out from the world's deodorant spray cans. Toluene, benzene, acetone and similarly toxic compounds are emitted freely in the manufacture of textiles, paints, and other goods which we must have for the good life. Smoke from the chimneys of industry, too, contains other components apart from the acid gases currently fashionable in the popular press. I expect that most readers will have heard at least a passing mention of the gradual destruction of the earth's ozone layer as a result of the pollutants we're producing, a destruction which is slowly removing the protection this layer affords from the potentially harmful cosmic radiation constantly bombarding the planet from outer space. In time, so the experts tell us, this radiation will cause a greatly increased death rate from cancer (if it's not curable by that time), a cheerful prospect but one which should effectively cut down the need for travel and hence air pollution!

Now we come to the question of water pollution. Just as the air we breathe must be pure enough for us to be able to get our oxygen from it, so also must the water we drink be pure enough to let us make use of this vital component of life without risk of poison. Unfortunately, mankind has only just begun to realise, once again, that the world's water supply isn't inexhaustible. For too long, we've carelessly thrown the refuse of our life-style into the rivers, lakes or oceans of the planet, heedless of what our thoughtless actions were doing to the precious water. You've only to visit the shores of the Mediterranean, where the beaches are stained from sewage washed up by the tide, or watch the dead fish float past you as you sit beside the Great Lakes, to have some sense of the damage being done to the waters of the earth. Walk along the beaches of almost any resort in the world and you'll find the brown patches of oil which mark the traces of a passing tanker weeks or months ago, or the green slime of algae spelling death to other living creatures in the sea or lake there. Try to drink the water in a quaint village in South America, Africa or Asia, if you insist, but do so at your peril and don't ask me to meet your medical bill when dysentery or worse grabs you.

There are so many ways to pollute water that it's almost unthinkable to expect pure water anywhere. Human excrement is allowed to flow untreated into pipes leading out to sea, so that we can have a rerun on the next tide. Chemicals of all types are distributed generously by manufacturers who find the local river just too convenient for words when it comes to disposing of the awful smell or colour, or the poisonous or corrosive by-product of their production process. Mercury levels in fish or in the creatures (including mankind!) who feed on them are a matter of grave (an apt expression!) concern even as I write these words.

The contents of a ship's waste container, or its spent fuel, are so easily tipped overboard when the vessel has left port. And why bother to repair the faulty exhaust system or leaking fuel line? After all, it's only a bit, and it goes to waste in the sea where it harms nobody. The home handyman who fixes his own car to save money doesn't need to worry about the residue from his oil change, does he? It's only a bucketful, and if he does the work over a drain it saves the bother of carrying the messy stuff elsewhere to get rid of it. In the meantime, his wife is dumping the remainder of the phosphate detergent into the same drain after finishing the week's laundry. There are more troublesome things than pirhanas being dropped into the subterranean waterways of the world.

Land pollution, too, is one of those insidious problems that are gradually creeping up on us to emerge unexpectedly, at some time in the future, and collect the debt we've accumulated over the years or centuries. We happily throw our fertilizers or pesticides or herbicides on to the ground, confident in the knowledge that we'll have a greener lawn, or better flowers, or fewer weeds and pests. We dig a deep hole and bury those unpleasant compounds, like the infamous PCB, which result from the manufacture of chemicals needed to cosset our modern way of life, or those radioactive substances, such as strontium, iodine, or lead isotopes, which are left over when the original material is no longer powerful enough to drive our generators or wreak havoc on our enemies. As long as we never disturb the unpleasant things we've dropped onto or in the ground, we needn't concern ourselves with their subsequent existence. So what if it rains? The chemicals will only get washed deeper undergound. What if they reach a spring, by some miracle? That's okay, because the water gets to the sea eventually, and that's where all the garbage is dumped anyway. Love Canal? Oh, yes, I heard of the place. It was in the news briefly some time ago, but that's all blown over now.

We just simply don't believe that the solid materials we throw to the ground can possibly ever come back to haunt us. They're not pollution, they're the necessities of life when our lawn has to be as green as the neighbours', and who wants to pay top price for electricity when they can make it cheaper by using radioactive sources of power? As long as it's buried in somebody else's backyard, of course. We all know what radiation sickness did for the inhabitants of Hiroshima and Nagasaki.

The short-term effects, ranging from instant death or vaporisation to milder (but much more painful) chemical changes that brought about a slow, lingering death are the ones usually mentioned in the anti-war propaganda, but there are other horrors still with us and likely to remain so for generations to come. For the survivors of the immediate blast, life held such attractions as stupor or incoherence for the few days before death, or painful, ugly and repulsive symptoms of burns that lasted for weeks before the victim eventually died. "Minor" problems such as loss of appetite, vomiting, nausea, prostration, watery or bloody diarrhoea, fever, loss of infection resistance or blood-clotting ability before the blessed release of a death delayed for months or even years affected thousands. Even the ones who still survive knew the agony of weeks of suffering, with some of these symptoms in a lesser form accompanied by loss of hair, premature ageing, or genetic mutations which we still can't quantify. In many cases, the onset of symptoms was delayed by several years, so that they're only now making their appearance.

Even with this awful evidence before our eyes, we haven't learned to treat radioactive substances with caution. It never ceases to amaze some people that the chemicals they dump on the land can reappear in the waters of the world and find their way ultimately into the water supply they depend on for survival. A child is born with hideous deformities, but we never connect that tragic event with the waste products we strewed so carelessly on the ground. The inhabitants of a village near a nuclear reactor exhibit an abnormally high rate of cancer, but there's no connection between the two observations, of course. We know that long-term effects of radiation include damage to growing cells, often passed on to succeeding generations, that can slow down their growth and so prevent division at a rate high enough for normal healing and human development to take place. Embryos fail to develop, adult skin refuses to heal, blood chemistry fails, eye opacity increases, and life span diminishes as a result. Yet, with all this evidence, radiation is still used to treat skin infections and x-rays are given indiscriminately, in many parts of the world, for shoe fittings or routine dental inspections.

The last of my pet hates in the realm of pollution is that of noise. Again, I find myself almost alone in my concern for this disquieting (if you'll pardon the pun) form of problem. If I go to a beach, or a picnic area, or a peaceful lakeshore, the last thing I want is the latest in rock music to keep me company. Yet, when I venture to protest, however mildly, I'm regarded as some kind of spoilsport who's trying to ruin somebody else's good time. After all, if I wanted to be a stick-in-the-mud, why would I come out into the country? The place to be miserable is at home by myself.

In the good old days, when technology was much simpler than it is now, a university friend and I built the most marvellous device that would eliminate the unwanted noise of radios for an appreciable distance around us. It was a kind of transmitter, tunable to any frequency, which would cause every radio within about two hundred yards to emit dismal howls when it was switched on. It was a boon to peaceful existence, and I've no doubt that it kept many a serviceman busy trying to locate the problem or fault in a portable set, but those days, alas, are no more. The advent of cassettes, FM and similar harbingers of interference-free reception has ruined the usefulness of my prize possession. I believe it was also illegal, but that never troubled us as we lay sunbathing in the quiet peacefulness of a summer afternoon!

Not that the radio is the only, or even the main, source of noise pollution, the emission of acoustic waves of a type and at a volume likely to cause distress in people exposed to it. The jet aircraft zooming over our heads, the cars or trucks roaring past us in the streets, the pneumatic drills tearing those same streets to pieces, and the factories turning out the conveniences of modern life all add to the din surrounding us. Worst of all, we've come to accept the increased noise level as an unavoidable part of life. Small wonder that our teenagers happily destroy their hearing by crowding round the local rock group! After all, they're only preparing themselves to be members of the modern, noisy world.

In one aspect of my research, I've been concerned for some time with the problem of noise in textile production. The problems had been around for over two hundred years, and several generations of deaf ex-employees have borne mute witness to its importance, but I'm still regarded as a trouble maker at times for crusading against it. It's impossible to convince a hardheaded manufacturer that the employees he's subjecting to unbearably loud noise are undergoing subtle psychological changes. The research evidence is meagre, but deafness appears to lead to isolation, excessive worry about health, depression, lack of interest in others, feelings of suspicion or persecution delusion, compulsive behaviour and obsessive thinking. In severe cases, there is an inability to understand or accept reality, an insatiable need for stimulation, excitement or action, and the frequent experience of tinnitis or vertigo. Small wonder that anti-social behavior, which has been shown to lead to reduced efficiency, greater absenteeism, negligence, carelessness, wilful damage, and a host of similar activities, is usually regarded as a likely result of deafness!

Fortunately, I've found a way to reach those employers who've no interest in the welfare of people working for them. I point out, quite simply, that noise is a form of energy. As such, it has to be supplied by the power generated (and paid for!) by the manufacturer. If he wants to continue pouring money out to make a big noise, that's fine, but there is an alternative if he is interested. When money is at stake, he usually is. Now, all I have to do is persuade the people who make trucks or aircraft or pneumatic drills to think the same way. Then I'll be ready to tackle the radio stations!

All of which brings us, not very satisfactorily, to a realisation of what man is doing to nature. Slowly but surely we're destroying the world around us. We're killing off the natural animal, vegetable and mineral life of the planet. We take the elements from the earth, use them carelessly, and throw them back in a changed form to damage irreparably the diminishing remains of the part we haven't yet touched. We modify the balance of species for our own selfish purposes, oblivious of the fact that we're creating untold and unforeseen havoc in the chain of life and that we ourselves are at the end of that chain. And that is where the chips start to fall in the wrong way, because Mother Nature can get her own back on us and is going to start doing so with a vengeance at any time now.

What we've so far, as a species, failed to realise is that we're paying, in our reduced comfort level, for what we're doing to nature. Worse still, comfort in the long run (as we've already seen) becomes tantamount to survival, and it's this factor which daily brings us nearer to Armageddon. So far, you see, our pollution has merely made us a little less comfortable. Of course it's unpleasant to have to look at unsightly litter strewn around the place, but what happens when the litter becomes too big to go away or when there's nowhere to store it any longer? When we begin to trip over it, or slip in it, or cope with the rats breeding in it, the purely psychological discomfort which bothers me (among some others, no doubt) will suddenly turn into a physical problem which has to be dealt with before it overwhelms us. If you find this hard to believe, think back to the last time you encountered, or read about, a large city when its refuse removal workers went on strike.

Similarly, the atmospheric pollution which gives us the psychological discomfort of fog can become a little more dangerous when the air's oxygen supply is so depleted that we're no longer able to breathe. When we've to fight the fish to get fresh water and the fish are already dead before we take the first sip, we may decide that the drink isn't quite to our taste (sorry about that one!). The land pollutants from our gardens or our power stations won't look so attractive when they're reappearing as poisons, or malignant tumours, or odd-shaped deformities on our mutant children. After the noise we've tolerated for years has finally made us into a race of deaf bellowers, perhaps we'll be able to appreciate the finer things of life as we "listen to" a Beethoven symphony on a cathode ray tube, watching the little squiggles jumping around because we can no longer hear the notes. "Comfort ye, comfort ye my people" may well be a despairing cry from the lost past in those halcyon days! I can only hope I'm not still around to see it.

CHAPTER 8

FASHION AND COMFORT

MAN'S psychological comfort has always, for as far back in history as we can trace, been closely connected with the clothing he used. In the earliest times, when "garments" were worn purely for decoration of the body, it's fairly certain that the finest of the miscellaneous bits of skins which our ancestors wound around themselves were reserved for the religious leaders or best fighters of the tribe. These people, on whom survival in the face of visible enemies or unseen spirits depended, had to be made to feel superior to the rest of the common members of the tribe who followed them into battle. As a mark of their exalted status, then, they were given the best skins, tails, or other assorted parts of slaughtered animals to wrap round their waist, head, or wherever else took their fancy.

When this custom evolved into the notion that status in other fields of life demanded recognition, the situation became more widespread. The garments worn by a member of the community identified him or her as being of a particular rank or station, so giving the wearer the mental satisfaction of being accorded the honour relevant to the position. Thus, psychological comfort, even in the earliest days of our race, was provided by the prevailing fashion in a very real manner. As time passed, the connection became stronger rather than diminished. Religious organisations, from the earliest spirit worshipers to the modern-day Pope or other church official, have always depended on garments to distinguish them from their flocks. Subtle differences in clothing, of colour or style, enable those familiar with the robes of office to recognise a cardinal, archbishop,

bishop, or other member of the ecclesiastical hierarchy at a glance. The same is true, though to a lesser extent, in the Anglican or nonconformist denominations.

Military leaders, similarly, are distinctive in appearance because of the insignia of rank which they wear on their uniforms. The shoulder badges worn by officers of the armed forces of any country in the world identify instantly (to somebody in the know!) what rank the wearer holds. To a considerable extent, there is a physical, as well as psychological, effect if you consider the different treatments enjoyed or endured by generals and gunners, respectively. Even if they're both unfortunate enough to be captured by the enemy, the general is usually accorded better treatment (at least in modern times) because of his rank and his increased importance to his captors. If the two soldiers were naked, however, they'd be indistinguishable to their foes, so the clothing they wear does indeed influence their comfort in other ways than the obvious one!

In the more mundane aspects of life, the same ideas prevail. The rich man, or his wife, must be identifiable as such to give him a sense of superiority over his poorer fellow-beings. The simplest way to get the message across to the greatest number of people in the shortest possible time is to flaunt a vast array of new clothes at regular intervals. The ability to maintain appearances and to be seen in the latest fashions in public is an indication that the wealthy are able to spend money without thought, to the envy of lesser mortals. This gives them the psychological boost, or comfort, which all human beings must acquire from some source or other. Unfortunately for the rich, the ability of mankind to mimic also comes into action at this point. Exclusive fashions are copied as soon as they appear (within hours in some cases) so that the masses can afford to be dressed like their financial betters, who must therefore buy another set of clothes to keep ahead of the game and remain looking different from their imitators.

And so it goes on. As soon as a fashion becomes too common, it must change to give the opportunity for conspicuous wealth to be displayed again. Hence, in our times, we're accustomed to seeing two, three or four fashion changes every year (at least in the Western world), to the despair of those who're concerned about the resulting effect on the earth's supplies of raw products or energy and the delight of those who make their fortunes by dictating what we'll wear for the next few weeks or months. The principle has always existed, let me say. It's only in the last few decades, though, that the accelerated frenzy we know today has become common. Until the advent of mass-production techniques, it was impossible for rapid changes in fashion to be dissipated at high speed, so you find that trends occurred much more slowly in past ages. If you stop for a moment to consider the fashionable clothes of the past few centuries, you can't help noticing how slowly changes occurred in early times and how much more rapid (and extreme!) they've been in the past few hundred years.

There is another sense in which fashion can assist us in achieving psychological comfort. At any given time, there are a range of garments available, from the most staid to the most outlandish, as defined by the ideas prevailing at the time. Within that range, each of us can choose to take a position which displays to the world our self-image. If we want to be thought weird or different, we can choose our clothing to convey that message to those around us. If we want to be regarded as conservative, on the other hand, we can dress accordingly and people will recognise the image we try to present. If we so wish, we can give both impressions on different occasions, so separating our social life from our working day, for example. In other words, the fashion we select will tell the onlookers something about our character and spare us the task of having to do so repeatedly to all we meet. This ability to project our own image of what we are is one of the most important functions of fashion in modern life and is of great assistance in helping us to achieve psychological comfort among those around us. On the street of any major Western city, you can readily identify which man is the city gent, the country squire, the working-class layabout, or any other of the stereotypes so familiar in our modern world.

Let us, then, examine this matter of fashion a little more closely. I confess, right at the outset, that I'm not an expert on fashion. I don't claim to have any knowledge whatsoever of the reasons why people will slavishly follow the decrees of a handful of men (usually the fashion dictators are male) in Paris, or New York, or wherever the current emperor of fashion happens to hold court. To be honest with you, I think the world might well be a saner place if we all began to follow the example of China in the sixties and ignored fashion. The need to run up huge bills would disappear, so making us all more relaxed (and less likely to have heart attacks when we work overtime to make a bit more money to pay off this month's accounts). Crime would be reduced. If you're not short of money, there's no need to rob somebody, and if we're all dressed alike, there's no way of telling who's worth robbing anyway. Add to that the fact that we'd not need to rob the earth of its minerals for the raw materials of textile production and you end up with a strong case for the abolition of fashion. Who cares if three Frenchmen, an American and a dozen wholesalers have to go on Social Security?

However, I don't expect the world will be clamouring at my door to become disciples of my new philosophy this week, so I may as well while away the time of waiting for it to happen by discussing the realities of fashion as they exist (temporarily, of course!) at present. The experts on the subject tell us that fashion follows cycles. If you examine the measurable trends (such as the length of a skirt or the width of a trouser leg) over a long period of time, you'll find they oscillate from one extreme to another. Consider, for example, the preferred length of skirt dictated by fashion during this century. The bottom hem oscillates in a fairly regular manner from ground level around the turn of the century to near-indecency at the height (literally) of the mini-skirt craze in the early seventies. As a result, it's possible in theory (given enough storage space) to follow fashion very inexpensively, if you so wish, by recycling your clothing every twenty or thirty years.

There are two reasons why this isn't a common practice. In the first case, we tend to think that, because we know the garment we're wearing is a quarter-century old, everybody else is aware of the fact, and this gives us a feeling of mental discomfort which prevents us from putting it on in public. Secondly, manufacturers and fashion designers aren't stupid, so they make sure the current fashion isn't an exact replica of the one prevailing last time the same ideas were in vogue. If skirt lengths are at knee-height this year, they decree that skirts will be flared (or will have some other variation that is instantly obvious) if they were tight last time they were at knee-length, and vice versa. There are always ways to ensure that the wearer of a garment can be made to feel psychologically uncomfortable if it doesn't belong to the current crop, and there are always enough people around to be the dupes of this policy. If we were all a little more self-confident, so our ego didn't need quite so much boosting, then the fashion world would grind down to a happier, less-frenzied pace and we'd stop allowing ourselves to be manipulated by remote dictators to whom we're totally unknown.

While the fashion craze exists, however, there is an even more powerful reason for us to follow it. This is the so-called peer-group influence, the effect of what our friends say about us and about our appearance, that compels us to buy, do, or say things we'd otherwise not consider. You've only to go into the high schools of your area, wherever you may live, to see the effect at work. Teenagers are very susceptible to the opinions of others, mainly because they're in the process of changing from children to adults and are unsure of themselves as a result. When a uniform is imposed on them by authority, an outcry not heard since the bricklayers at Jericho worked to rule arises on all sides. Let them choose their own clothes, though, and they all wear exactly the same, even to the brand name of the jeans they buy, because "everybody's wearing these this year." The same is true, in a more subtle way, of their elders. Check the clothes being worn next time you go to a formal banquet or a sporting event. You'll find that the two sets of clothing will be very different from each other, but that all the articles at either of the two functions will be of similar type.

Peer-group influence isn't purely a phenomenon of the Western world, or of the rich, either. If you visit skid row in any big city or if you go to the far corners of the globe, the same principle is at work. Rich or poor follow fashion of some kind. All the members of a given society, whether that society is the millionaire beach set in California or the Bedouin tribes of North Africa, will be dressed in a similar manner. There are obviously certain local factors which will influence the choice of styles, but, once these are taken into account, there will still be considerable scope for variation. The fact that it doesn't exist to any appreciable extent is a sure indication that the pressure of the society is at work in influencing the freedom of choice. Psychological comfort needs the approval of society, and if the fashion dictates of the society are ignored, that comfort is denied.

Perhaps the most noticeable example of the way in which fashion is followed relates to the style of garment worn. Not only must the length of the hem be correct but also the way in which the article is designed. It's no use wearing a dress of exactly the right length if the style isn't the current one. The addition of a frill, or a bow, or a particular type of sleeve is enough to destroy the usefulness of a dress if no one else is wearing the same modification. You've only to examine the illustrations in a book describing this year's fashions and compare it with last year's to understand what I mean. Slight changes will have been made, even if major ones haven't taken place, to make the appearance subtly different. Again, the omission of these changes would lead to such psychological discomfort that the wearer would literally not be able to appear in public in the garment. If you don't believe me, look at comparable pages taken from fashion catalogues a year apart. The differences will speak for themselves!

Until the most recent times, it was mainly the ladies who were so averse to being thought old-fashioned, but the malady has spread to the male members of the human species, at least in the West, within the last twenty years or so. As a result, it's not uncommon nowadays to find people who regularly discard perfectly serviceable clothing and buy an entire new wardrobe full of garments on a regular basis. As you may have gathered by the tone of this chapter, I'm not among those who feel it necessary to throw away a suit on the way home from the tailor's and I tend to wear clothes until they become unacceptable for physical, rather than psychological, reasons. If only the rest of the world would follow my shining example, life would be so much more pleasant and comfortable for all of us!

The next thorny aspect of fashion which influences comfort is the question of colour. As I'm sure you're all aware, it's utterly impossible to be seen in a public place in a garment of unsuitable colour. Unless the shade of green matches your face or the shade of red your bloodshot eyes, the world will regard you as a pariah. The reasons for this are purely psychological. There are colour combinations which arouse feelings of comfort or discomfort in the brain; some combinations are normally regarded as harmonious and others as discordant. If the latter happen to be associated with an individual's clothing, a vague sensation of unease is generated within the mind of an onlooker. So, for instance, a pink dress worn by a lady with red hair "clashes," as we say, in the opinion of an observer. Brown shoes worn with a grey suit, or black ones with a brown suit, somehow look out of place.

Colour is a part of the mechanism by which we express our personality, as discussed above. For an extrovert, bright colours, or shades which are deliberately chosen so that they clash, can announce to the world his (or her) outgoing nature. For the respectable businessman, a sober shade of grey can perform the equivalent task. Colour can influence the way in which people think about us, for we all have a preconceived notion of what looks good on everybody else we meet. It's somewhat surprising, in a way, that there is a reasonable consistency in our opinions. Agreement about what is a suitable colour combination to coordinate a particular type of individual is quite widespread. A person who looks exactly right to one onlooker can often expect to get the same reaction from the majority of observers. Similarly, somebody who looks "wrong" to one person will probably look the same to most observers. There really is a direct connection with the clothing worn; you can put exactly the same outfit on two different people (of the same size, so that judgement isn't clouded by poorly fitting garments) and feel that one person looks just right while the other has a totally unacceptable appearance.

The universality of agreement on colour suitability must be the result of some deep psychological property of the brain. Possibly the frequencies of complementary colours produce some kind of electronic harmony in the mind when light waves set up the same kind of discordant effect as do two musical notes which are off-key. The true physical reason, as far as I'm aware, has never been investigated, but perhaps the ancients were ahead of their time when they talked about the harmony of the spheres!

The last aspect of colour I want to discuss is that of uniformity. Dyeing skills are such that a fabric isn't exactly the same colour at all parts, but the differences in depth of shade resulting from a local small difference in concentration of the dye molecules are (in a successfully dyed material) indistinguishable to an observer. Tests show that people don't all have the same ability to see variations in shade, and it's therefore possible that one person may reject as unevenly coloured a material which looks perfectly uniform to somebody else.

Even so, the manufacturer must either achieve extremely good uniformity or be prepared to risk possible poor sales of his product because it doesn't meet the critical standards of some potential buyers.

To a greater degree, the same reason can cause some individuals to wear clothing which is psychologically uncomfortable for most of those who see it. A form of colour blindness is responsible for the problem in both cases. In its milder manifestation, it prevents discrimination between similar shades, and in its more serious occurrences (unless a deliberate attempt to create a particular effect is being made), we assume the wearer has a brain totally unable to register colour disharmony as defined by the majority opinion.

Now for a comfort-related aspect of fashion which affects physical as well as psychological factors: the fit of a garment. Constriction, reducing blood flow, breathing ease, or convective cooling may be tolerated because physical comfort is incompatible with the psychological discomfort of "fitting wrongly." The clothing we wear can be made to surround the body in a range of degrees of tightness, from a loose, flowing style to a skin-strangling contour fit, and each of these types is acceptable under the right conditions.

There are many examples of garments in which the ability to drape or cling is an important factor in the design of the clothes. Long, voluminous evening dresses would be far less attractive if they stuck out stiffly from the body instead of flowing down from it. Conversely, a pair of jeans which flowed round the wearer in a comparable manner would be rejected as hideous by their embarrassed owner.

Psychologically, again, peer-group influence is significant in making the decision on how close the fit should be. If we're considering a teenager's jeans, current fashion decrees that the body should appear to have been poured into the garments, preferably so efficiently that they'll never be removable without recourse to a pair of scissors (or so it would seem). On the other hand, very few Arabs wear skin tight robes, even as teenagers. It's partly a matter of cultural tradition, usually governed by external factors (the need to emphasise body contours or to stay cool in a hot desert, respectively, in these two cases), but also the influence of the society in which the person lives. An Arab teenager who has been assimilated into a Western culture wears his jeans every bit as tight as does his locally born friend.

This brings me to the final two points of this chapter, the effect of ethnic traditions and the compromise between different types of comfort. The act of moving from one culture to another can be a traumatic experience, as I can witness from personal memories associated with emigration from England to Canada. One of the ways in which the original ties are maintained is by retaining the style of clothing used in the homeland, and the prevalence of ethnic dance groups, etc., across the North American continent typifies this attitude. On the other hand, one of the quickest ways to find acceptance in a new land is to adopt as many of the customs as possible, including that of garment style, as soon as you can. The Westernisation of many people from Oriental parts of the world is the obvious example of this trend. In other parts of the world (whether East or West), of course, styles are very different from those we're accustomed to seeing in our own country, wherever that may be. Half of the pleasure of travelling is to see at first hand the different customs of other lands, and it's the clothing worn there which is the most obviously distinctive feature of many countries.

Figure 32 shows some national costumes worn by various races around the world, and you're invited to while away a train or plane journey trying to identify them all! In the proper setting each one of them looks just right, but out of context they all look ridiculous. Picture alpine or Balinese dancers performing in business suits or cocktail dresses, and think whether you'd then travel halfway around the world to watch them. Yet, these (to our eyes) strange styles of garment are the accepted normal modes of dress in those far-off places. These are the clothes which society in the particular country favours and which the people must wear in order to feel comfortable in that society.

One important aspect of clothing comfort often neglected is that of functionality. In other words, does the garment do what it's supposed to do? There is not much point in wearing a raincoat that is the most fashionable and attractive garment in the world if it won't keep out the rain. You may be the envy of everybody who sees you in the coat, but if you're soaked underneath it, the envy is small comfort. Besides looking good, clothing must perform well and do all that is reasonably expected of it if the wearer is to be truly comfort-

Figure 32. National costumes of various regions of the world.

able. The occupational clothing worn for horse riding, space exploration, fire fighting, football, or any similar use where protection of the body is needed would look cumbersome or ridiculous in the wrong setting, but the comfort, or even survival, of the wearer (or somebody dependent on him) would be at risk in its absence.

As a next step, then, we should consider how fashion influences the total comfort of human beings, a matter which sheds rather a different light on the subject. It's an undisputable fact that some of the clothing styles dictated by fashion trends are not, by any stretch of the imagination, comfortable in the physiological or physical sense. Consider, to name but a few examples, the corset or bustle worn by women around the turn of the century, the stiletto heels of two decades ago, or the tight jeans of today's teenagers.

Nobody in his (or her) right mind would choose to wear these garments if the vagaries of fashion didn't compel such folly. The bustle must have been murderously uncomfortable to sit in. (I'm afraid

I can't speak from personal experience, as I'm too young to have worn one, not to mention being of the wrong sex.) The corset was roundly condemned by medical opinion of the time for its tendency to constrict the thorax and cause health problems. Stiletto heels were notoriously dangerous in their tendency to do unpleasant things to the ladies' feet and ankles, and tight jeans must rank as the world's third-worst strangulation device after garrottes and boa constrictors. More to the point, the latter garments also bring about additional discomfort in the form of dermatitis and yeast infections at unmentionable body sites. Why, then, do people wear these outrageous articles?

The answer, I think, lies in the dominance which psychological comfort enjoys over the other two kinds. We're so afraid of the opinions of other people about us that we're prepared to go to virtually any lengths to win their approval. If fashion says we must sport clothing we'd shun like poison, given our freedom of choice, then sport it we will. We're prepared to sacrifice our bodily comfort for the sake of our mental comfort, even at the risk of literally harming ourselves physically in the process. As long as we meet the conventions, we subjugate our real preferences and pretend the garments we wear are ideal. Ask any modern teenager (or the strange older people who would like to be thought of as teenagers) if a pair of tight jeans is comfortable, and you'll get the same kind of response as you'd have got from their grandmothers or their mothers. Of course they are! The bustle or corset gives a woman support, high heels improve the balance, and tight jeans . . . well, I've not heard any really convincing excuse yet, but I'm sure there is a perfectly logical explanation of why they're so comfortable!

All of which brings us to the realisation that somewhere in the world there are people, just like you and me, who're aiding and abetting the international confidence trick perpetrated on the poor unsuspecting public by the demons of fashion. The sequence of events, I'm told by the experts, is the same for all fashions in all walks of life everywhere on earth. A new fashion is announced and worn by the models of the top designers. The style is adopted by one or two members of the idle rich, who flaunt it for a brief period of time. It's then copied by the manufacturers who make clothes for the mass market. At that point, it may either begin to become a craze or sink into obscurity. If the former happens, the fashion is adopted by the so-called fashion leaders of a particular district (town, suburb, street, etc.), the people who pride themselves on being avant-garde in their dress. The rest of the population then follow, like sheep, in sequence over the next few months until only a few of the ultra-conservatives are left. Some of these eventually succumb but only as the fashion is waning, so they join the bandwagon just as it's disappearing into the sunset to make room for the next trend. The whole cycle then repeats itself *ad infinitum* with each new fashion in turn.

It's interesting to identify the position on this fashion scale occupied by your acquaintances. If you look carefully and keep a record of all the clothing worn by your neighbours, you'll eventually discern a pattern. Make a note of when each type of garment appears in your neighbourhood and who wears it, then record the progress of each fashion in turn. There will be a consistent order of adoption of each one by the people you're charting, and the leaders or stragglers will become obvious. If you include your own name on the list, the situation becomes even more interesting, though you've to be careful not to bias your own buying habits as a result of your observations. I can pretty well tell you who the fashion leaders in my own district are, and I know I'm fairly well down at the end of the followers. In fact, I often miss out on a fashion completely, either because I don't like it or because I'm too idle to rush out to do the necessary shopping. So far, I've managed to live with the psychological discomfort which this is supposed to produce in me, though my wife suffers appallingly at the thought of what people's opinion of me must be. Give me physical comfort and a healthy bank account every time!

CHAPTER 9

COMFORT IN THE HOME

HOME is where we spend a major part of our life, the place where we're supposed to relax and let the day's cares slowly fade away after the work is done. All too often, in these modern times, home turns out to be the place where we have marital fights or work long hours into the night on the alterations, the children's homework, or the file we brought back from the office because the boss said it had to be ready for tomorrow's meeting. Home in the traditional sense, though, is the place where we most expect to find (and most need) comfort. What are the factors of importance in providing it? Let's consider them in turn.

I suppose the first need is for psychological comfort. If that exists, then all other aspects will follow or can be found. When the home, as is so often the case, is a hotbed of tension, then even the most luxuriously appointed mansion in the world won't seem comfortable. Mental comfort has to be provided by the people who share the home, and they must work in harmony together to ensure that it flourishes. I'll be saying a little more on this point in a later chapter. When you walk into your house, all the cares of the outside world should stay outside or the house isn't a true home in the sense of the word. Mental comfort implies the absence, or insignificance, of major sources of stress, such as financial, marital, health, parenting, safety, security, or similar ones. It's small wonder, in these times, that homes are no longer as comfortable psychologically as they used to be in bygone days.

Given that the need for mental comfort may not be met as adequately as we'd like, what else can be improved to make the goal of comfort a little more attainable? Possibly the most important aid is the provision of enough space, of the right size and design, for the inhabitants to enjoy adequate privacy without sacrificing social interaction. In other words, the house must be big enough (or designed well enough) for an individual to be able to find a quiet corner when the need arises but not so huge that he can never find anybody when companionship is needed. The definition of space needs tends to be governed by the society in which we live; a house which was easily large enough for my family's needs in England would be far too cramped for us after living in Canada for several years. Similarly, an Oriental society would condition its inhabitants to prefer more crowded conditions than an English resident could tolerate.

Although I've never seen any work to support the idea, I've got a sneaking suspicion that the space needs of a home are closely tied in with the optimum interpersonal distance of the members of a particular society. Certainly, I've noticed that the preferred interpersonal distance appears, in general, to be greater in Canada than in England, and greater there than in India, and the house sizes (or the space allocated to each individual within a household, which is a more relevant measure) appear to correlate well with this order.

The changes taking place in Victorian England, too, can be directly related to the matter of space. When the relative affluence of middle-class families allowed individual bedrooms for family members for the first time in history, privacy needs became expanded to the point where prudery resulted, and we all know how the Victorians are regarded in that respect! In modern-day North America, the urgent "need" for multiple bathrooms has been thought by some unkind behavioural scientists to be related to social status, rather than for the cleanliness which we all know to be the real reason behind this necessary expense for the poor householders. There is no proof for all these theories, of course; they're just hypotheses, a figment of my imagination. Some enterprising social scientist will no doubt be able to pull in a sizable research grant to demonstrate their validity (and get some very pleasant travelling donc at the same time!) after reading this book.

The designer of a house, then, should ensure that the volume of space is appropriate for the people who'll live in it and take care to see that the aesthetics are acceptable to those people. This can involve physical as well as mental comfort. It's no use expecting a six-foot man to be comfortable in a home with low doors, where he must duck his head every time he moves from one room to another. We've three hanging lamps in our home, which were the bane of my existence until I persuaded my wife to let me put a ten-foot-long dining table under two of them and a sideboard under the other. Now that I'm no longer in danger of imminent decapitation, I feel much more comfortable in my home and suffer fewer headaches (though whether that is from the reduced collision rate or the lessened tension as a result of a diminished need to be on the watch for violent blows, I wouldn't like to hazard a guess!). Certainly I'm able to relax, and thus be more comfortable psychologically, because I'm not constantly on my guard against violent blows on the cranium.

The actual design of the house can provide psychological comfort in other ways. Is each room planned efficiently so that awkward or frustrating movements aren't necessary to use it? A room should be capable of being lived in. The space must be large enough to let people move around freely, but not so huge that they have to walk a mile to set the table or turn on the television. It should be aesthetically pleasing, yet offer some degree of utility as well as beauty. Architects with a pet axe to grind often forget these simple facts and impose their own fanciful ideas on a client without any regard for the basic needs of the people who must live with their avant-garde theories for the next few years. You may have noticed, incidentally, that this chapter contains no illustrations. The omission is deliberate; there are so many different individual opinions on precisely what type of colour scheme or furnishing (or other such items to be discussed in the chapter) constitutes a good design for comfort in the home that I hesitate to inflict yet another one (my own) on the poor reader! There are, of course, some general rules that can be applied, and we can usefully consider each comfort-related aspect of the home in turn.

First, is the colour scheme a pleasant one, and does it blend well with the furnishings chosen? Are the walls overcrowded with ornaments that constantly need dusting, or are they so bare that the place feels like a prison? In either case, comfort is hindered, and something should be done to improve the situation. Is the kitchen too far from the dining area? Is it too cramped or so huge that the meal gets cold before you can carry it to the door? Is the place insulated properly? Are the lights (and their switches) in a convenient place? Is the sleeping accomodation large enough and in a suitable area, near enough to the rest of the house to reach before you fall asleep on the journey? Conversely, is it so near that any parties tend to take place on top of the blankets where you're slumbering?

Are the conveniences convenient, or just a primitive bit of plumbing at the bottom of the garden? Is there enough storage to let you keep the place reasonably tidy without having to rent a nearby warehouse? Does the house have all the advantages demanded as minimal by the society in which you're living, or do you feel that it's inadequate to represent your life-style properly? All of these, and many other, factors have to be considered before you can decide whether a place will truly meet your needs and, hence, enable you to feel psychologically comfortable in it.

The colour aspect can be critically important in achieving, improving, or maintaining comfort. Just as was the case for clothing, colours which clash can give rise to subtle feelings of discomfort in the mind, and this factor is of particular importance in the home, because it's not usually possible to do a quick change of the wallpaper or paint for the evening if it happens to displease you in any way. Colour harmony, and colours which suit the personality of the inhabitants of the house, must be selected with comfort in mind. Don't expect your sweet, eighty-year-old granny to feel comfortable if you paint her room a vivid mixture of psychedelic green, orange and purple. In the same way, don't give your motor-cycle-loving teenage son pink wallpaper with tiny roses all over it if you want him to retain some vestige of his sanity, let alone the respect of his peers. For total comfort of the entire family, colours within each room, and colours of adjoining rooms where they can be seen simultaneously, should harmonise pleasantly with each other and with the personalities of those people who spend a lot of time in them. If you must have your grandmother and your teenage son under the same roof, try to put at least a bathroom (which you can safely paint white) between them!

Now we come to the problem of what to put into the home. The function of furniture has changed considerably over the centuries, and the changes have all related in some way to an improvement in comfort. An easy chair used to be designed to allow people to sit upright, so that dignity could be preserved, and motionless, to avoid dislodging carefully applied wigs or powder. We recognise today that people are likely to adopt all kinds of peculiar postures, from sitting to reclining to lying to curling up in a ball, often in the same piece of furniture, so we give them a beanbag chair. Cupboards in a bedroom, so convenient for holding clothes or linens, originally served as a storehouse for food, so that the pangs of hunger on a cold night could be assuaged without leaving the warmth of a less-than-cosy bed. Feasting (and drinking!) were often the only consolation in a freezing bedroom far from the sole fire burning in some remote room elsewhere in the house. For the same reason, beds were often enclosed by curtains, not for appearance as in modern designs, but to cut down on the draughts and thus give marginally increased comfort.

A table was formerly placed near the wall so that guests at mediaeval banquets could eat from it by sitting on the bench fashioned as a part of the wall. The invention of chairs enabled people to sit round a table, instead of along one side, so giving an opportunity to enjoy less stilted conversation and thus enhancing psychological comfort. The use of chairs as status symbols should also not be forgotten. Ancient rulers who sat on an elevated throne, surrounded by their vassals, are no longer the inevitable rule, but let a newcomer to the house dare to sit in father's favourite chair and watch the sparks fly! In Oriental societies, the chair has, in many cases, disappeared completely, and cushions are used to emphasize the equality of all those who are in the house.

Thus, furniture will obviously be a major component in your scheme of filling a house to impart comfort to it. The design of the furniture, like that of the dwelling area itself, will have to match the personality of the occupants as far as possible, but there is also a more important physical aspect this time. Furniture has to be used by human beings, and they won't be inclined to use it if it's not acceptable from the point of view of their physical comfort. A chair is no use at all if it can't be sat in without discomfort, no matter how attractive its design. Even the bed slept in by King Egbert on the memorable night after he and Boadicea had met William the Conqueror and King John for a game of bowls on Plymouth Hoe, or something along those lines, isn't of any real value (except as an antique, in purely monetary terms) in a home where you want to have a comfortable bed. In the first place, it's probably two feet too short, because kings were built on the small side in those days, and in the second case the woodworm renders it in imminent danger of collapse at any moment.

So furniture, like houses, must be designed to be functional as well as aesthetically attractive. A good designer of modern pieces will use ergonomic principles in his work. That is, he will take measurements on human bodies, in whatever posture is suited to his needs, and design what he is making so that physical comfort is enhanced by the shape of the article in question. One of the latest fads, the beanbag chair already mentioned, is, you might say, self-designed, because it takes up the shape of the person sitting in it and conforms to his or her body contours, no matter how peculiar the posture adopted. The same is true for a water bed, another recent innovation.

For more conventional items, the adjustment isn't quite so simple. It's not possible to design a piece of formal furniture in such a way that it will mould to any body placed in it, so the designer has to use his judgment, in conjunction with statistical tables of mean and variance values of human dimensions, to assist him. These are published, from time to time, in many countries and show the dimensions of average human beings of the country at the particular time. By reference to them, it's possible to decide what variation is permissible in the design of an article so that it will allow most of the potential buyers to use it without feeling uncomfortable. The aim is to meet the needs of as wide a cross-section of the population as possible, and the extent to which this is achieved is a measure of the degree of acceptance which the design will eventually enjoy (on the assumption that other factors, like aesthetics, don't render it unacceptable).

The other types of furnishing in a home will be chosen on similar grounds. The carpets will be of a colour and style to suit the individuals living with them, but will also be installed in such a way that they are physically comfortable. Underpadding can add a cushioning effect, making sure that the level of foot comfort is acceptable as you walk around the house. If there's too much padding, however, you may well find yourself in dire risk of twisting an ankle as you sink into the soft ground with every step you take. The choice of fitted versus area carpeting has to be made with both mental and physical comfort in mind. A fitted carpet may be preferred for ease of maintenance, but an area rug will undoubtedly show to best advantage the beautiful polished wood flooring which was put in at great expense when the house was new. Remember, though, that the wood has to be kept in a polished state of shininess and, apart from the work involved, the chance of slipping may thereby be increased.

Cushions enter into the design of a comfortable room in any house. As usual, they must match, or complement, the rest of the room in colour without being too obtrusive, yet they mustn't inhibit a comfortable sitting position. If the place is full of well-stuffed cushions, it may not be possible to find somewhere to rest without a lump constantly pressing into your back. Personally, I always remove any cushions surreptitiously from behind my back, when I'm sitting in a room where they've run amok, and slide them on to the floor or under the settee. My wife gets a little concerned about what the hostess might think of me, but at least I'm comfortable while I'm there. Anyway, I can always hope the dog gets blamed for my misdeeds, and besides, I'm not sure I want to be invited back to a place where they feel that their cushions are sacred.

Curtains provide further scope for improving the overall comfort level. Apart from their purely decorative effect, they fulfil an important function as barriers to heat, cold, or excess light. A curtain eliminates the stark lines at the edge of a window frame, making the transition from window to wall softer, and hence more comfortable, to the eye. As an additional bonus, they also act as thermal insulators because of the air spaces which they enclose, a matter we've already discussed in an earlier chapter. If the external temperature is greatly different from that prevailing inside the house, drawing the curtains can prevent the transfer of large quantities of heat through the window. Glass is a bad insulator, so heat can flow readily through it, and the presence of an additional heat barrier in the form of a curtain can improve considerably the temperature comfort inside the room (or the size of the bill needed to keep it at a satisfactory level).

Bed linens are particularly important in enhancing comfort. The human body tends to have a lowering of temperature during rest and has to be kept insulated to compensate for the reduced production of heat if a comfortable night's sleep is to be enjoyed. Sheets and blankets provide the necessary barrier to heat loss, so that a cold feeling doesn't waken you from your slumbers. If ever the bed covers inadvertently fall off on a chilly night, it's virtually impossible to remain asleep, and totally impossible to fall asleep again after you wake, until you've restored things to their normal state. I once made the mistake of camping out on New Year's Eve in a tent with no form of heating inside it. Even though I crammed every bit of clothing I could find on to every portion of my body, stuck my feet into a rucksack and my head under the supposedly warm sleeping bag, I was still frozen. I can still recall the agonies of that long night before I could safely emerge and try to revive my congealed bloodstream.

There is also another, perhaps equally important, function of bedding, and that is to satisfy the need for tactile comfort. The sheet, which has the task of isolating the blankets from the body so they don't have to be washed quite so often, mustn't be rough to the touch, nor must it be so soft that it tickles. In my various sojourns in the beds of the world (none of which was ever inhabited at the time by a lady other than my wife, I hasten to add), I've experienced the extremes of softness and hardness, from the scratchy ferocity of a potato sack, as I recall it, which I was loaned at a youth hostel in my younger days, to the slippery softness of a smooth nylon sheet (in a North American hotel which shall remain unnamed) on which I slid neatly from the bed immediately after jumping into it. The ideal sheeting fabric is one which feels soft to the touch, yet doesn't make you think you're about to be hurled out on to the floor if you turn over in your sleep.

Table linens, too, can enhance comfort. The cloth on which a meal is served can add mental enjoyment to the act of dining if it's clean, crisp and stain-free. Physically, it shouldn't be rough to the touch, nor should it cause the cutlery to snag or slide as it's used or laid down. Colour and texture, as usual, should be pleasingly unobtrusive to enhance the decor of the room and the dishes set out for the meal. The serviette or table napkin should obviously match the cloth and should, in addition, be smooth so that no abrasion can create the discomfort of scratchiness as you wipe your mouth. The linen towels used to serve or dry the dishes shouldn't shed lint over everything they contact and should be able to remove water (or the gravy, if that is what they are for) efficiently and without leaving smears of dampness or stickiness behind.

The last type of soft furnishing I want to consider is the other kind of towel that is used to dry the miscellaneous bits of human beings rather than the dishes from which they eat. It's even more vital for comfort that a hand, face, or bath towel should be absorbent, soft and luxurious. There is nothing quite like the sensation you get when a warm towel enfolds you as you emerge from a hot bath. On the other hand, there is nothing worse than finding out that it merely slides across your skin without removing a drop of water, leaving you with a sinking feeling as you realise you'll have to face the cold world with clammy skin and clothes. In winter, the resulting discomfort can be appalling. The textile factors involved in imparting all these properties, as of all the other furnishing goods I've mentioned in this chapter, will be discussed in Chapter 12 of this book, but it's important to recognise at this stage that there are methods for making any soft article of this type comfortable.

Next we must consider three factors of great importance in the search for comfort in the home: the levels of sound, light and heat which are available. Mental and physical comfort are both affected by each of these factors. It's impossible to feel comfortable psychologically if you're not comfortable physically, as I've tried to stress several times in this book. Bodily and mental comfort are so intricately bound together that we simply can't isolate either aspect totally without giving some consideration to the other. Thus, in the home, the noise, the brightness and the temperature must all be set at a level where we can be relaxed in the mind as well as unstresed in the physical body.

In the case of sound level, there are two factors to consider. The noise must be at an intensity which doesn't harm us medically and must be of a type that we don't find displeasing. Deafening noises from outside the house, caused perhaps by passing vehicles, voices raised in a shout, or the sound of an engine on the lawnmower, snowblower or road repair equipment, should be reduced in volume by the acoustic insulation used in the construction of the building.

Noises arising either inside or outside the house don't need to be deafening, though, to be annoying and hence uncomfortable. If you happen to be a rock fan, the sound of a Beethoven symphony can be anathema to your tender ears. If, on the other hand, you've undergone the metamorphosis of age and become a civilised human being with an ear for good music, then the cacophony blasting out from your teenager's record player in a sealed room half a block away can drive you into a frenzied fury even when it's almost at the point of inaudibility (by the teenager's definition, that is). In other words, the sounds reaching your ears should be at a moderately low level and of a kind not displeasing in their nature.

The total absence of noise, incidentally, can be more uncomfortable than you might expect. Anechoic chambers (rooms with no echo and which therefore sound absolutely "dead") are very distressing to anybody placed in one for any length of time. I've used one for my acoustic work from time to time and can't stand to be in there for more than a few minutes at once. Similarly, people who find themselves in a totally silent environment often begin to talk to themselves. Ask a housewife, alone in the home all day, why the radio or television is kept on, and she'll explain that it's "for company." Mention of the environment brings to mind the importance which, as usual, it has in this case. The sound of birds twittering is a blissful one when you're out in the country for a pleasant walk, composing a brilliant piece of poetry about spring, but the same birds screeching outside the window as you wrestle with insomnia, the income tax forms, or the essay that has to be finished within the next half hour can be a little less idyllic.

Light may also be a mixed blessing. The level of lighting which you need for an intimate party would cause untold eyestrain if you tried to read by it. Conversely, the light which you need for comfortably scanning your page would destroy any illusions of romance if you examined by its harsh glare a member of the opposite sex who was potentially to become an intimate friend (and you can put your own interpretation on the expression!). There are established levels of lighting to suit the needs of each type of situation, and it's wise to abide by them if you want either your eyesight or your love life to last for more than a brief fraction of its normal duration. The direction of light, too, can be critical. The light from the sun which dazzles you at breakfast is a nuisance, but the same intensity of light coming from over your left shoulder may be exactly what you need when you're threading a needle for sewing or carrying out some equally intricate work. In general, a good lighting system should provide adequate illumination for the task in hand but shouldn't create discomfort by shining into anybody's eyes.

Finally, we come to the question of heating a house. There is probably no other factor quite as important to so many people as the maintenance of a comfortable temperature in the home. If the sound or lighting level is uncomfortable, we can always turn a switch to change the situation instantly. In addition, we're not usually making use of the sound or light throughout every minute of every day. With the heating system, it's a bit different. We're constantly trying to maintain our internal temperature at its desired level, as we saw earlier in the book, and every bit of help we can get improves our comfort. We're also exposed to the ambient temperature for the entire duration of our stay in it, without respite, so must constantly be coping with whatever temperature setting exists.

In times of moderate season, that is probably not a difficult task and may well be taken care of without our noticing anything, but in extreme conditions life isn't quite so simple. If the external climate is very cold or very hot, then the provision of supplementary heating or cooling is critical to our comfort and to our survival, possibly, in the case of a cold environment. Try to imagine life in the northern parts of North America, Europe or Asia without the benefits of winter fuel. Read the accounts of racial tension in the large American cities in summer, when the temperature soars and the poorer inhabitants lack the blessings of air-conditioned comfort. It soon becomes very clear that, at least in modern times and in the Western hemisphere of our planet, there are people who can't survive in sanity and comfort without the presence of some form of assistance in the form of auxiliary heating or cooling.

It's clear, then, that in the home, as elsewhere, comfort is a necessary attribute of modern living. A combination of design and technology has made possible the life-styles we enjoy and has raised the appreciation of comfort to heights unknown only a few short years ago. We have to remember that, as a result of the tremendous strides made in recent decades, it's possible for us to enjoy, in the place where it's most vital, the mental and physical comfort which we so desperately need.

One difficult question needs to asked, though. Have we, in the constant striving for comfort, been damaged as a race by losing some of our ability to cope with change in our environment? This ability has brought mankind to his present state of achievement, against all the odds that faced our remote ancestors when they emerged from the trees in that far-off era. Have we sacrificed it in our softness, and will we suffer the consequences at the hands of nature, just as the Romans suffered the consequence of their softness at the hands of the barbarians? Only time can tell; maybe the more hardy members of our species will be the sole survivors of the splendour that is today's human race. Blessed are the meek indeed, for they shall inherit what's left of the earth. These are disturbing questions, questions that can't be answered anywhere, least of all in a book devoted to human comfort; it gives me a sense of discomfort even to think about them!

CHAPTER 10

COMFORT IN WORKING LIFE

HAVING ESTABLISHED the factors affecting our comfort in the home environment, we can now discuss the way in which the workplace can be influential in modifying how comfortable we feel. For most of us, our work occupies at least as much time as we spend at home (not counting the time spent in sleep at either place, that is) and often more. Thus, perhaps one-third of our adult life is spent at some kind of working situation, whether at an office, a factory, a laboratory, in the home, or maybe in a vehicle. I'll be dealing with transportation comfort in the next chapter, but the other job locatons are all relevant to our consideration of how comfort level can be maintained or increased during our working hours.

Possibly the easiest situation to consider, since we've already made a start on the subject, is the one where work is carried out in the home. The worker may be a housewife, performing the important and often under-esteemed task of ensuring the continuation of our species and society, or a self-employed person working from the home, or somebody (like a writer) who requires no special place in which to earn money and so can use the home as an office. In all these cases, there'll be different needs, and we've already discussed some of them in the last chapter. For all of the people involved, some measure of privacy is of even greater importance than is usual in a household environment. A housewife needs time to herself, away from the stresses of being with the children constantly. The self-employed worker must be able to rely on peace to carry out the task of bookkeeping or other needs of his professional life. A writer can't expect to produce good work in public. Thus, in all cases, privacy is vital, though for different reasons and possibly to different extents.

This entails the need for adequate space. It's no use trying to get complete privacy in one corner of a room where others are likely to be carrying out tasks that interfere with concentration. For a routine purpose, such as reading a book for relaxation, that kind of "privacy" is perfectly acceptable, but if long hours of working time are involved, a totally private room, closed off from all other activities, is a necessity. It should preferably be reserved entirely for the purpose of work, since other things have a tendency to creep in unnoticed if you relax your vigilance and allow it to become a dual-purpose place. As I look around my own study, I find all kinds of distractions have been deposited here, from books to chess sets to art supplies to rock collections to microscopes to ornaments (and worst of all!) to the decanter of sherry to tempt me away from the straight and narrow!

Putting all thoughts of these temptations nobly aside (and throwing away the key to the booze cupboard), I press on with the task in hand. In terms of physical comfort, I'd suggest there are three areas of particular importance, these being those of location, functional efficiency, and ambient conditions. Location should be chosen for relative isolation from the remainder of the house so that minimum interference takes place. There is little future in an attempt to work in the room next door to the kitchen, where the neighbours or their children drop in to meet or talk with other members of your own family. My own study is on the top floor of the house, well out of earshot of potential interruptions, and if I close my door I can't even hear the house doorbell ring. That's luxury!

By functional efficiency, I mean the provision of a working environment which allows the task to be performed in a safe manner, with no distractions, without appreciable waste of time or effort. The layout of the working space, with no dangerous lamps looming just above the skull or slippery mats lying in wait for an unwary step, is an obvious example of this aspect of the need. Chairs should be comfortable enough to be sat in for long periods of time without backache but not so comfortable that you fall asleep the minute you get within six feet of them. Tables, desks or benches should be at the right height so that you can work on them without climbing a ladder or looking like the Hunchback of Notre Dame. Equipment and supplies should be within easy reach without too much effort but not tumbling all over the working surface so you can't see beyond the end of your nose. If the work involves a lot of movement (as for the housewife, for instance), then the layout of the area should be planned well to minimise the distance which must be covered between individual tasks.

Ambient conditions perhaps cause the most difficulty in terms of identifying them suitably. The temperature, lighting and sound levels are so much a matter of individual preference that it's virtually impossible to give advice on how they should be set. All that can be done is to provide information for you to use in making your own choice. We've already seen what latitude is permissible, without involving the body in undue stress that can be physically harmful. Now it's a matter of identifying the optimum levels (with some idea of individual variations that can be tolerated) at which human beings function.

For temperature, the answer is relatively simple. A well-nourished human body functions at its maximum efficiency with an environment in the range between sixteen and twenty degrees Celsius. At lower temperatures, concentration can be hampered by the need to provide extra heat, usually by shivering, and at higher ones lassitude sets in because the thermoregulatory system (assuming normal clothing is worn) is quiescent, so blood flow to the body (and hence to the brain) becomes sluggish. Unfortunately, we've become conditioned to demand higher temperatures, in general, during the cold weather of winter, and few of the homes in North America are ever allowed to cool down to an efficient temperature. Possibly the fuel shortage which we're assured is almost upon us will prove a blessing in disguise by allowing us to develop more brain activity as a side benefit. Certainly, there are homes in Canada where such an incentive to reduce temperature is desperately needed!

Sound and light levels are a little more difficult to tackle, since they don't have quite as strong a physiological basis on which to provide guidance regardng optimum levels. The usual recommendations, derived by psychological studies and physiological measurements, are that the ambient sound level should lie somewhere between 40 and 60 dB(A), and the lighting level should be set in the range 40 to 100 foot-candles, depending on the activity taking place. At lower sound levels, the quietness can begin to become oppressive, and at higher ones distracting or damaging. At lower or higher lighting levels, eyestrain can occur, either by peering to see clearly or by squinting against the brightness. Having said that, it's necessary to add the rider that some people are totally unable to work within these arbitrary limits. I know many who must have a deafening row in the room before they can shut out all other distractions, and I know others who can't possibly concentrate unless the proverbial pin dropping (and that's even smaller than a mouse dropping!) sounds like a thunderstorm. Lighting levels, too, can be varied to suit individual idiosyncrasies well outside the range quoted, though I must confess that I can't honestly remember anybody with unusual preferences in this area.

Now we move out of the home to look at other working environments. All the factors I've discussed so far in this chapter are relevant, though possibly in modified form, in other work situations. There, though, they're so complicated by the fact that there are usually several people, all with individual preferences, to be considered. Let's consider an office, for instance. The usual picture brought to mind is of a room, large or small, with desks, filing cabinets, typewriters, probably a computer or two these days, furniture of greater or lesser luxury and, of course, people. If the office is that of the chief executive, it will usually be for his exclusive use, but for lower ranks shared space is normal.

In modern, open-plan facilities, where the entire staff of a section is located, it's common to make use of partitions as separators to give some semblance of privacy. Now, what are the comfort needs in these conditions? We should take a closer look at the occupants. In the hypothetical office, there may be one or more supervisory staff members with subordinate personnel in the same area, and their needs may be different. The supervisor will probably be involved mainly with paperwork requiring concentration, while the other occupants will be engaged in tedious or repetitive tasks which can be made more pleasant by conversation. Some of the people may need soft music to make their lives tolerable, while others at the ends of the acoustic taste spectrum will only be comfortable if rock music or silence, respectively, is provided.

The other comfort-related aspects of the home environment are still applicable here, however. The functionality of the furniture is a matter that can be arranged, within reason, on an individual basis. As long as the chair a staff member wants isn't exorbitantly priced (and, more to the point, doesn't have higher status than his superior's!), there is no valid reason why he shouldn't be asked for his preference. The fact that he's consulted can only lead to improved relations, and hence efficiency, in the company.

Safety, from the aspect of low obstructions, slippery floors or projecting desk corners, must still be an important consideration. The planning of the office to minimise unnecessary movement between desks or sections can make life more comfortable for the occupants. Lighting levels can be made reasonably personal by the skilful use of spot or desk lamps. Temperature, though, is a difficult matter. In the present state of technology, it's virtually impossible to maintain a variety of different temperatures within the same enclosed space, because convection currents soon ensure that equalisation takes place. Possibly some future genius will invent the perfect combination of a space heater and microwave oven, so that a field of localised energy can be tapped only by the person at whom it's aimed, but those days aren't yet in sight, mainly because of the dangers associated with the radiation of microwave energy.

It's in the realm of industrial settings that the provision of comfort becomes a more serious problem. It's a little difficult to generalise about the facilities existing here. Industrial surroundings can vary from an environment closely resembling an office to one looking, sounding and feeling like Dante's Inferno. Hence, it's impossible to stipulate exact rules for the provision of comfort, which acquires a different (and purely relative!) meaning in specific circumstances.

In the first case, there are more people involved, with all the difficulties surrounding individual tastes compounded. Then there is the fact that the space requirements are different. Usually a large room, in which many people must move around to a greater extent, is associated with the industrial setting. Finally, the production process itself may complicate the situation. Light industry can be carried out in relatively pleasant surroundings, but the customary vision that springs to mind isn't at all pleasant, including as it does any or all of such nastinesses as high noise levels, dirty air, hot or cold temperatures, and potentially hazardous working conditions. A steel mill, for instance, or an open-air lumberyard are unlikely to be quiet places, and, in winter, the two will certainly be at opposite ends of the furthest possible deviation from optimum temperatures or comfort. The difficulties are further complicated by the matter of cost. It simply isn't possible to heat, light or soundproof a large industrial building to any satisfactory degree at an economic rate. All that can be done is, at best, a rather poor compromise in which the minimum necessary modifications are made where absolutely essential, and the remainder of the plant is merely treated to make it safe without being comfortable.

This mention of safety, the critical need to prevent employees from being exposed to the risk of accident, disease, radiation, noise, or other potentially harmful agents, brings us to the point that accident prevention, a crucial factor in the area of comfort, is often a more serious matter in industry than in the office. Industrial hazards frequently include such dangers as high voltage, hot substances, dangerous chemicals, toxic solvents or vapours, high places, radiation, high-speed machinery, or similar sources of peril. In modern industrial planning, accident prevention has become a prime area of importance, fortunately, since the progress of technology has multiplied the risks faced by workers in recent decades. Educational programmes designed to increase the awareness of the worker concerning the dangers lying in wait have improved the safety, and hence comfort of both a physical and mental kind, prevailing in the factory.

One of the important aspects of accident prevention lies in the equipment used. In today's industrial processes, the twin requirements of complex operations and high production rates give rise to a need for highly sophisticated machinery, with many moving parts, possibly operating at high pressure, temperature or voltage. Consequently, safeguards must be more carefully considered. Is a machine safe? Does it have any unguarded sharp edges, moving parts, or electrical components which could prove harmful or fatal? Are there any auxiliary parts that could trip, hit, or impede an employee operating it? Is it likely to leak a poisonous or otherwise harmful chemical, a fuel which will ignite, or steam to scald its operative? Are the power supplies and wiring safely hidden behind ducting, or are they exposed so that spills of oil can destroy the insulation? Are there any water leaks, or poorly drained areas, so that footing can become unsafe? I'm sorry to say I've been in plants where all of these perils were present, and you have my assurance that I wasn't the least comfortable, mentally or physically, in any of them.

Unfortunately, we tend to become blasé about danger because it's not manly to be afraid, so the risks are all too often disregarded. The government inspector, if he happens to notice them (and I've known plenty of cases where blatant dangers were completely overlooked on inspection), is regarded as a snooper who's trying to justify his existence at the expense of the poor manufacturer. This mental attitude is one of the most dangerous aspects in any plan to improve safety and comfort in the factory. I must have been in a dozen or more plants where the wearing of hearing protection was compulsory, because of high noise levels, and where the employees refused to use their earplugs because they felt "soft" for doing so. Their toughness only begins to exact a toll many years later, when it's too late to reverse the inevitable brain or hearing damage cause by their mistaken "macho" attitude. There's ample evidence, as is illustrated by Figure 13 in Chapter 5, to show that hearing damage results from exposure to excessive noise. Hearing loss progressively increases from moderate (30-45 dB), in which conversation is difficult and anxiety from other causes increases, to profound (80-100 dB), in which only very close speech or loud noise can be distinguished, often only with the aid of amplification.

At intermediate stages, conversation becomes increasingly more difficult without the aid of amplification and the sense of isolation gets steadily more pronounced. Social and environmental contact both decline relentlessly, identification feelings are impeded, and an ever-increasing dependence on visual or tactile senses evolves in an attempt to compensate for the lost hearing. It's even possible to demonstrate a detectable loss of hearing during the passage of a single shift, a procedure which, it's hoped, will serve as a dire warning to employees reluctant to use protective devices. The employer, in many cases, is at least partly to blame. If an order was enforced strictly, with warnings and dismissals for disobedience, the problem would become much less serious, but employers themselves are often resistant to legislation designed to protect workers and condone the disregard of safety precautions.

Eventually, it's the manufacturers (and their customers) who've got to pay for this attitude. Discomfort can lead to job dissatisfaction or to increased absenteeism. Efficiency diminishes, and wilful damage can be caused. Deliberate reduction of effort is one way to wreak revenge on a featureless job that's uncomfortable both mentally and physically. Carelessness about the quality of the product or about the conditions under which it's produced can lead respectively to a loss of reputation and hence of work or a loss of health, with the same end result. In serious cases, disability can be permanent or life itself can be sacrificed; the psychological discomfort isn't diminished but merely transferred to somebody else. Wilful damage can not only slow down production but can also mean large repair bills. Even if the damage is through neglect, rather than by actively destroying machinery, the cost is still the same. Unless workers, trades unions and management cooperate to provide as much comfort as possible in the working environment, the cost of a disgruntled worker will never be avoided.

In the long run, this cost is borne by society. Higher repair bills mean that a product must be given a higher price. If the consumer has to pay more, his income won't stretch as far and he'll buy fewer goods. If he buys fewer goods, the unemployment rate will rise. When that happens, government assistance costs, in the form of unemployment benefits, have to rise. As a consequence, tax increases occur, thus reducing still further the buying power of each consumer and compounding the problem.

It is, unfortunately, only in difficult economic times that the realisation of the truth of this inevitable vicious circle strikes home. When it does, it's too late to do anything about it, because the men who caused the situation in the first place are numbered among the unemployed and can no longer change their ways. So, too, are their innocent colleagues who had no hand in the consequences of dissatisfaction. It's a sad fact that when full employment resumes, it's only a matter of time before the lessons of history are forgotten once more and the vicious circle begins all over again. The blame can't be allocated squarely to any one segment of society. Employers must share with workers the responsibility of providing an attractive work environment. Government must do its part in making the legislation reasonable and giving assistance in meeting it. Society must give a clear indication to its leaders that it expects true concern for the problems of the workplace. Only then will we be able to hope for harmony (and thus comfort) in the working world.

CHAPTER 11

COMFORT IN TRANSPORTATION

MANKIND has become a race of travellers. Whether for business or pleasure, a fair proportion of this globe's inhabitants spend a considerable part of their life in moving from one place to another. The movement may be in the company's private jet, by public transportation, on foot, pushing a barrow or guiding a beast of burden as escape is sought from the horrors of persecution or war. From the very rich to the abysmally poor, human beings can be found undertaking journeys at all levels of existence. In their travels, they try to be as comfortable as is feasible under the circumstances. In this chapter, I want to examine the factors which contribute to this comfort.

Perhaps we can start at the most physical aspects and work our way along the path from physical to mental ones until we come to the truly psychological areas. By this scheme, the first topic to be considered is the way in which the transport method adopted affects the human body undergoing it. We can begin, since most methods of travel involve the act of sitting or being supported in some way, by dealing with posture. The body's skeletal structure is intended to maintain the frame in any of several positions so that activity is possible, but only a limited number of them can be adopted for any length of time if comfort is to be satisfactory.

As you'll realise from personal experience, there are essentially only lying and sitting positions which can be assumed for comfortably passing the hours needed to complete a plane, train, bus or similar journey. If we're lying down (as in a sleeper, for example), then the surface beneath us has to be reasonably soft. If we're sitting, the same is true, with the additional need for softness yet firmness in the back support of the seat. In a seated position, too, we've got to be able to extend our legs and sit back into our seat. The Victorian ideal of a straight-backed chair is of little use when any distance must be traversed. Travel is an exhausting business, and an upright posture for the extended time taken would lead to quite severe aches in the neck, back and legs during and after the journey.

There is also a need to change your posture occasionally to avoid becoming cramped in one position. When we stand, sit or lie in a fixed position, our bodies (unless we happen to be astronauts in free space at the time) are subjected to the influence of the earth's gravitational field. Pressure of the blood increases below the level of the heart and is reduced in the parts of the body which are higher than that particular organ. Resulting effects can include tissue swelling, pooling of the blood, and changes in the shape of blood vessels. If the position is maintained, circulation to the body may be reduced to the point where nerve impulses are prevented from being transmitted, and the body responds with an attack of cramp or other warning signal to tell us to adjust our posture. No matter how comfortable the bed or chair may be, we need to turn over (in a prone state), adjust our exact position (when seated), or even get up to walk around for a change from either a lying or sitting location.

It's important also to remember that a sudden move, especially of large magnitude, can produce discomfort. As we saw in Chapter 5, the act of standing from a prone position causes about half a litre of blood (the equivalent of an appreciable haemhorrhage) to rush suddenly from the chest to the feet. Vasoconstriction to reduce pooling as much as possible is an automatic reflex action, but its effect takes several seconds to restore normal blood flow conditions and, during this time, blood pressure in the brain drops dramatically. Faintness and darkening of vision occur, and any heart defects can bring about severe problems. This kind of change may be present, or even magnified, in flying and is of particular importance when you call to mind the reduced atmospheric pressure adopted on modern aircraft.

As a result of all these concerns, a great deal of effort is devoted to the task of designing the seats for air, train, bus or car travel, all of it aimed at the ideal of a support which remains comfortable over long periods of time and under all foreseeable conditions of use. The most recent ideas favour a seat of the type shown schematically in Figure 33, in which the body can maintain a relaxed posture, with legs extended but feet touching the ground, and with support being given at the sides as well as at the back. Even with this seat, however, there'll be a need for occasional standing or walking excursions if the passenger wants to be free from stiffness at the end of the journey.

One point to be made is that the requirements for seat design will depend to a considerable extent on what its occupant is supposed to be doing. The design of a car seat for the driver, for example, should be different from that for his passengers, since they have different needs. The driver must be comfortably alert, whereas the passengers should be able to relax during the travelling period. Very few, if any, modern cars take account of this difference, so that the seats are usually a compromise between the two requirements, with a tendency to lean towards the driver's needs for reasons of safety. This makes sense, since a driver who falls asleep isn't of much use to his passengers, so they are expected to suffer a little inconvenience for the sake of their survival chances. Perhaps the matter of individual seat design for the different positions in a car should be examined a little more carefully by the automobile manufacturers.

Next we come to the matter of motion and the way in which it affects comfort. Once the passenger is settled into a seat, the vehicle can depart and its movement must be considered. One of the most severe problems involved in transportation design is that of vibration. When movement takes place along the ground (or through the air), the vehicle is subjected to numerous small shocks in a continuous manner as a result of multiple collisions with the road or the buffeting of air resistance. In addition, if the vehicle is a powered one, the engine will produce movements to contribute to the overall shaking effect.

Because of all this movement, the passengers will be constantly subjected to the forces of small amplitude which we call vibration. Each solid object present in the vehicle has a tendency to move at a natural frequency and, when it does so, can oscillate about a mean position as a result of the kinetic energy supplied to it from its surroundings. The net result, as experienced by the passenger, is a series of small-amplitude movements, at a fairly high (and very

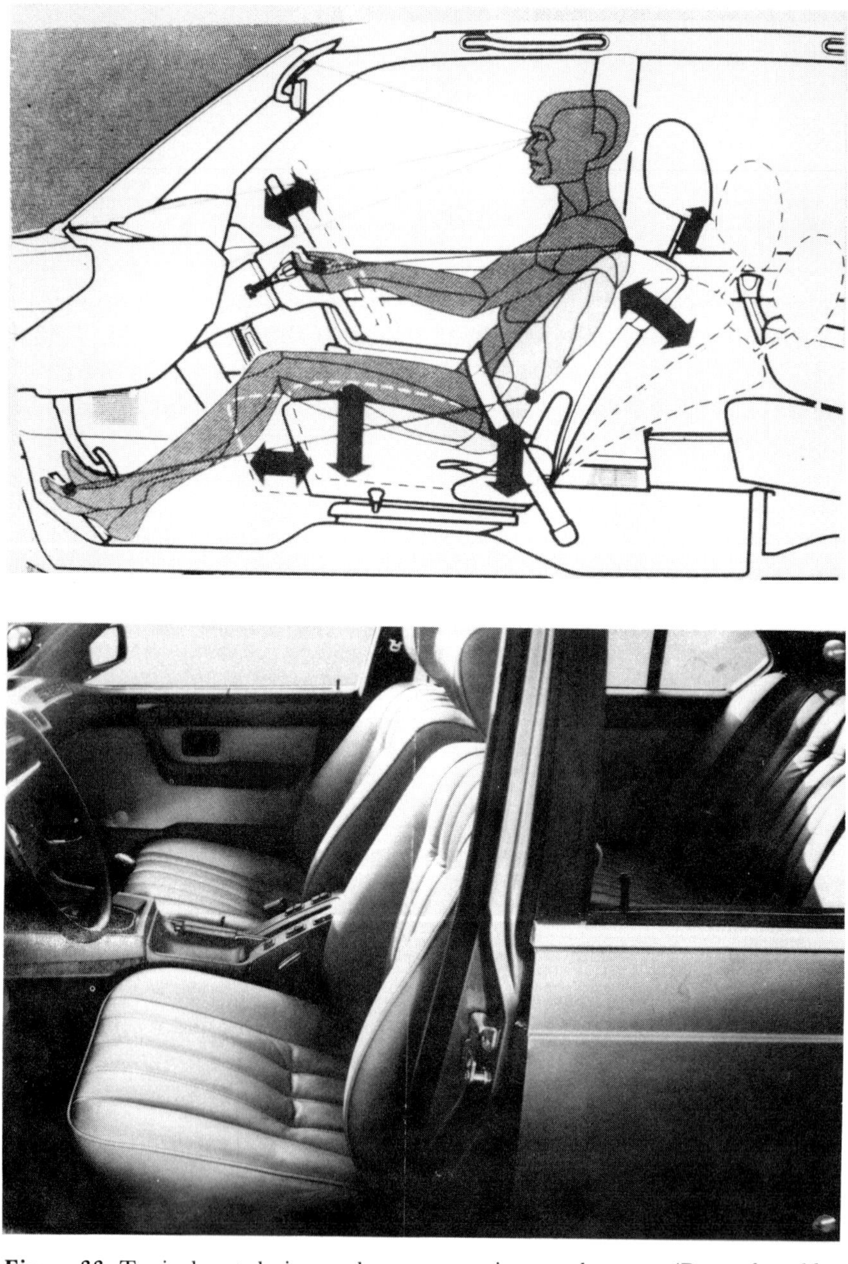

Figure 33. Typical seat design and appearance in a modern car. (Reproduced by kind permission of BMW AG).

unpleasant) frequency, that shake his entire body and produce severely uncomfortable effects of sickness, headache, fatigue, or disorientation. If no precautions are taken to eliminate the movements of the vehicle components in contact with the passenger, the ride will be an extremely uncomfortable one.

The body doesn't like to be shaken too much. The internal organs can respond by functioning at reduced efficiency or by failure if the motion is too great. The brain, thrown round inside the skull, can suffer damage in severe cases. Muscle ache can become troublesome. At the very least, the passengers are liable to suffer travel sickness because of the unsettling effect of the motion on the balance mechanism within their inner ears.

The effects of vibration are removed essentially by a process of damping, the reduction in amplitude of an oscillation by absorption of the energy it contains. Two approaches are feasible. The first of these is to reduce the incidence of movement before it's caused, a process accomplished, for example, by tires on a wheel or by using plastic (rather than metal) components in the moving parts of a vibration source. The second is to interpose flexible mounting or coupling components in the transmission path along which the vibration travels on its route to the passenger, as exemplified in the use of shock absorbers or springs. Figure 34 illustrates typical shock absorbers of the spring and hydraulic types, commonly used to make sure that the passengers in a modern car are isolated mechanically from the vibration induced as the vehicle moves over an uneven surface. Damping components, such as rubber cushions, steel springs, or pistons in a cylnder, are clearly identifiable.

The first of these two possible ways of reducing vibration—the treatment at its source—is to be preferred if at all possible. As for a sound wave, prevention tends to be more economical, both in terms of power wasted and in reducing the frequency of repair or replacement. There is, though, only a limited number of ways in which you can reduce unwanted movement at the source. It may not be technically feasible to use plastic gears, for instance, because they are softer than hardened metal. In consequence, most of the work that is intended to improve comfort by cutting down vibration tends to be of the second kind.

The elimination of vibration between its origin and the passenger

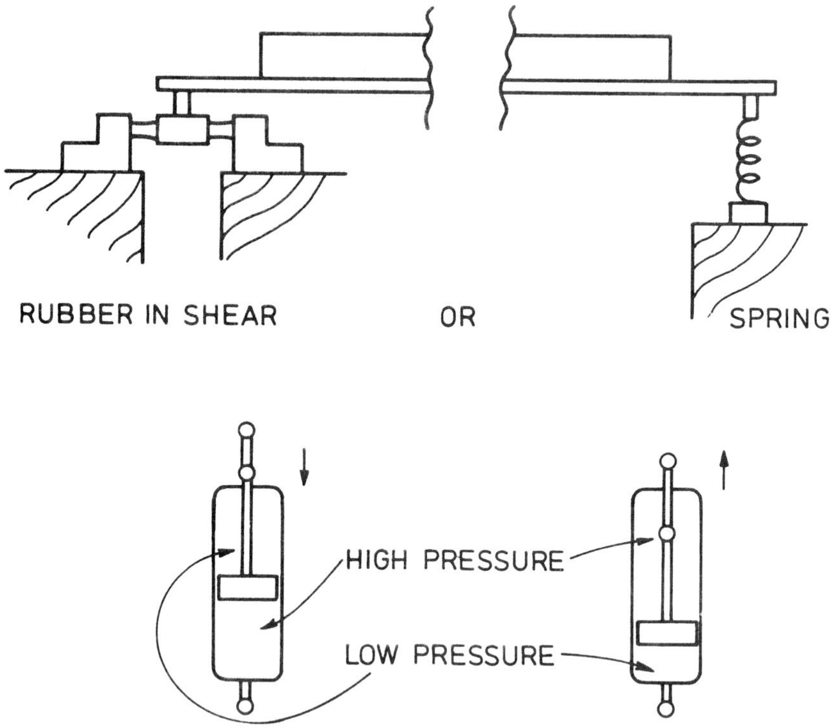

Figure 34. Diagram illustrating spring and hydraulic shock absorbers.

location is an area of work that has received considerable attention. The needs are slightly different, depending on the mode of transportation, since the forces encountered tend to differ. Aircraft trundling along the ground at supersonic speeds are about as rare as flying trains, and the types of vibration produced by high-speed travel along the ground and in the air are obviously not going to be the same.

Once the vibration has entered the vehicle, however, it makes no difference where it originated; it must be removed from the structure as thoroughly as possible, and this elimination follows similar principles no matter what the vehicle. The obvious starting place to tackle the problem, then, is where it first appears. For a ground surface vehicle, this typically involves the use of springs to isolate the wheels from the remainder of the structure, coil or leaf springs normally being incorporated into the axle system. Figure 35 illustrates

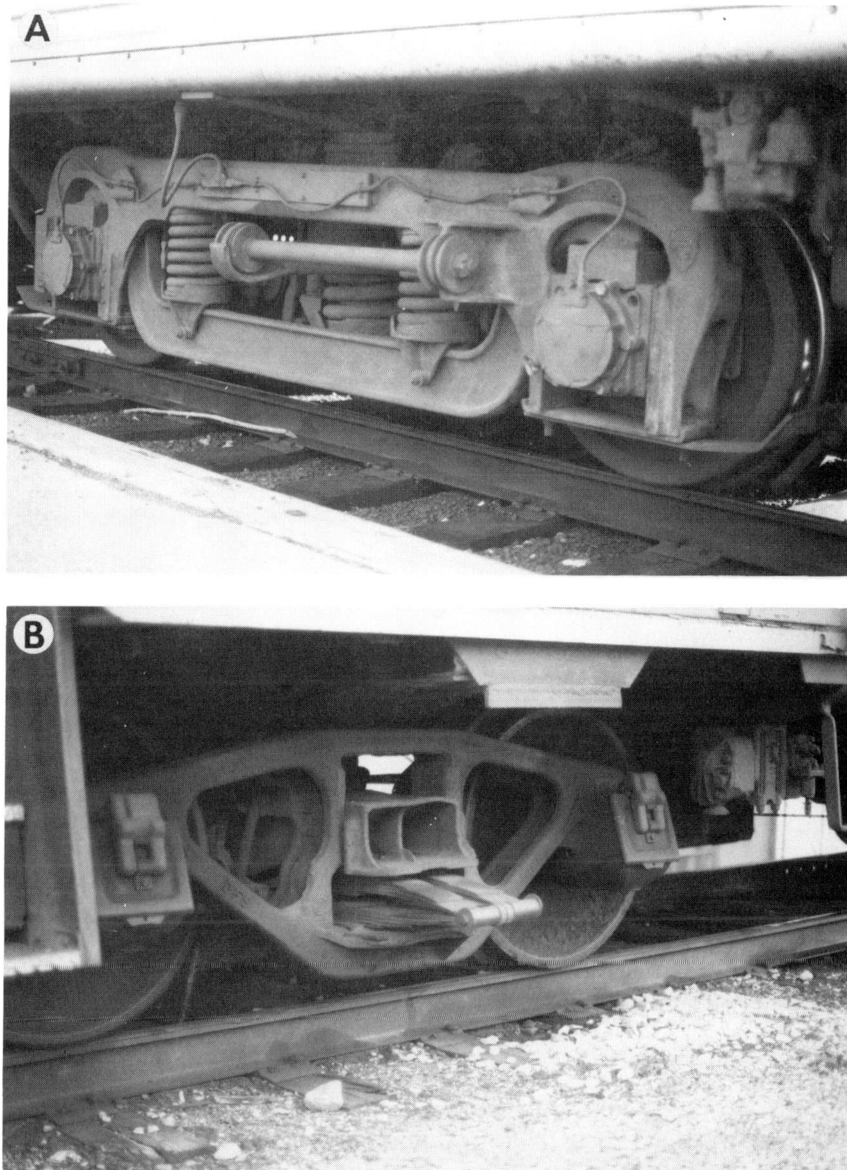

Figure 35. Springs in railway rolling stock: (A) coil type and (B) leaf type.

the use of springs in a railway coach, both types being visible in the diagram. A considerable reduction in vibration is achieved by this one step, and there is usually not much more done until the immediate vicinity of the passenger is treated.

In an aircraft, apart from the springs used to absorb energy during taxiing, little treatment is necessary, since the use of springs is pointless when airborne flight is in progress, but the aerofoil design includes a consideration of airflow over the surface of the aircraft to minimise buffeting as well as the adverse effects of drag. The designer of a modern aircraft wing section must make sure the shape is the best possible one to improve airflow at high speeds. In a ship, similarly, the aim is to reduce the impact of waves as well as to improve speed. The designer takes great care, for example, in establishing the shape of a racing yacht's hull to meet these twin requirements.

At the passenger position, any residual vibration is reduced as far as possible to the point where it hopefully isn't noticeable. The seat (or bed) on which the passenger travels is isolated from the remainder of the vehicle by springs, once again, normally only of the coil type in this case. In addition, an energy-absorbent layer of material, usually of an air-filled or foam construction in modern seats, is incorporated into the structure to give further cushioning from the effects of vibration. The damping of this arrangement is usually sufficient to remove all but the most severe effects of motion in any vehicle, so that passengers aren't conscious of being shaken around at all during the voyage.

One particularly troublesome type of vibration, however, which doesn't respond to treatment by damping, is the extremely low-frequency one as encountered in ships at sea. In severe weather, the ship may roll or pitch once every five to ten seconds or so, with the well-known effects on passengers unlucky enough to be prone to seasickness. Because the vibrations are not high-frequency ones, there is no high-speed movement, so damping and shock absorption are ineffective. The worst motions (in the pitching and heaving modes) occur when the waves are between three-quarters and one-and-one-quarter times the length of the ship, causing greatest displacement and hence greatest risk of breaking up. For travel sickness, though, the most familiar type of motion is rolling, which is again at a maximum when the period of movement is the same as that of the waves.

The method normally used to diminish seasickness risk, with only partial success, is stabilisation of the ship to prevent or reduce the most severe movements set up in the first place. A technique which has been in use for centuries is to make the keel as deep and as long as possible. Nowadays, anti-rolling fins and tank stabilisers are fitted to most oceangoing ships to reduce roll. A modern liner can be stabilised, by means of mechanical fins which are automatically deployed from the hull by a servomechanism to resist the rolling motion, to reduce the roll to about 15 percent of its initial size, but this may still be severe enough to cause motion sickness problems for poor sea travellers.

Now we must consider one of the most crucial factors (at least in my mind) governing comfort in transportation: the subject of ventilation. I can't speak for other people, but I personally can never be happy if I'm forced to travel in an enclosed space where I can't breathe freely. If I'm shut in completely (or even if I only think I am), I tend to become extremely travel sick. There is no more miserable feeling in all of human existence, I'm convinced, than the one experienced when you want to get out of a stuffy moving vehicle but can't. Two of my own pet hates are the smell of cigarette smoke and the smell of diesel fumes when I'm being jolted around in a bus, car, or any other means of rapid translation from place to place. I get very angry if I'm denied a non-smoking seat on an aircraft or if the smoking area is divided down the centre (a ludicrous arrangement practised by some airlines) so that smoke can waft freely across to me. I also become very queasy, if I travel by sea, when the waves are high and the weather is so bad that I have to remain indoors with the smell of diesel engine fumes in my nostrils.

The only way to avoid such unpleasantness (short of becoming a recluse and staying at home!) is to have adequate ventilation in any vehicle in which you must travel. Open windows or air-conditioning equipment are all of enormous help in combating the discomfort of travel sickness, but there are problems with these solutions. Windows tend to let in rain as well as air, air conditioning is expensive, and air lines are draughty. Also, air conditioning may be too space-consuming to put in any small vehicle.

Modern cars, trains, buses and other transportation devices are designed with air-exchange arrangements so that air can be drawn in from outside (often through a filter to remove particles) and allowed to flow freely but gently throughout the vehicle. Figure 36 illustrates, by a schematic diagram of the entire system and a photograph of the controls, the ventilation system used in a modern car. By its use, the air inside the vehicle is exchanged completely over a period of time and any tendency for it to become stale is reduced or eliminated. Breathing comfort is thus enhanced and the discomfort of a headache from carbon dioxide buildup avoided.

Mention of headache brings me to the next in my series of factors important in the transportation comfort field: the matter of stress. Stress can be of two kinds (mental or physical) and both have some influence on the sensations experienced by a passenger or driver. There is the stress of having to adopt an uncomfortable position in the journey, a factor which is clearly closely allied to the question of posture. There is also the psychological aspect of stress, the more usual one to be concerned about in travel. If a passenger is under mental strain during a journey, comfort can't be at an acceptably high level. Stress causes the body to be tense, thus aggravating the situation by bringing on a feeling of muscular tiredness, headache, cramp, or other unpleasant symptoms. Hence, stress levels are diminished as far as possible by the people responsible for the transportation arrangements.

Physical stress is reduced by careful design of seating, as discussed earlier, but psychological stress is a little more difficult to deal with. Everything possible is done to make sure the passengers aren't mentally upset. Soft music is played in an aircraft before and during each takeoff and landing to deaden the clunking sounds accompanying the plane's climbing or descending operations. Alcoholic beverages are dispensed continually to help fortify the nerves during a flight longer than a couple of hours.

I once had the unpleasant experience of flying with an airline (which I won't name) that was notorious for the poor condition of its aircraft. The interior was filled with loose or badly fitting sheets of metal on every bit of wall, floor, or roof surface, and the whole thing vibrated constantly with a harsh clattering noise. At least we weren't bothered by any peculiar sounds from outside, because they (and the

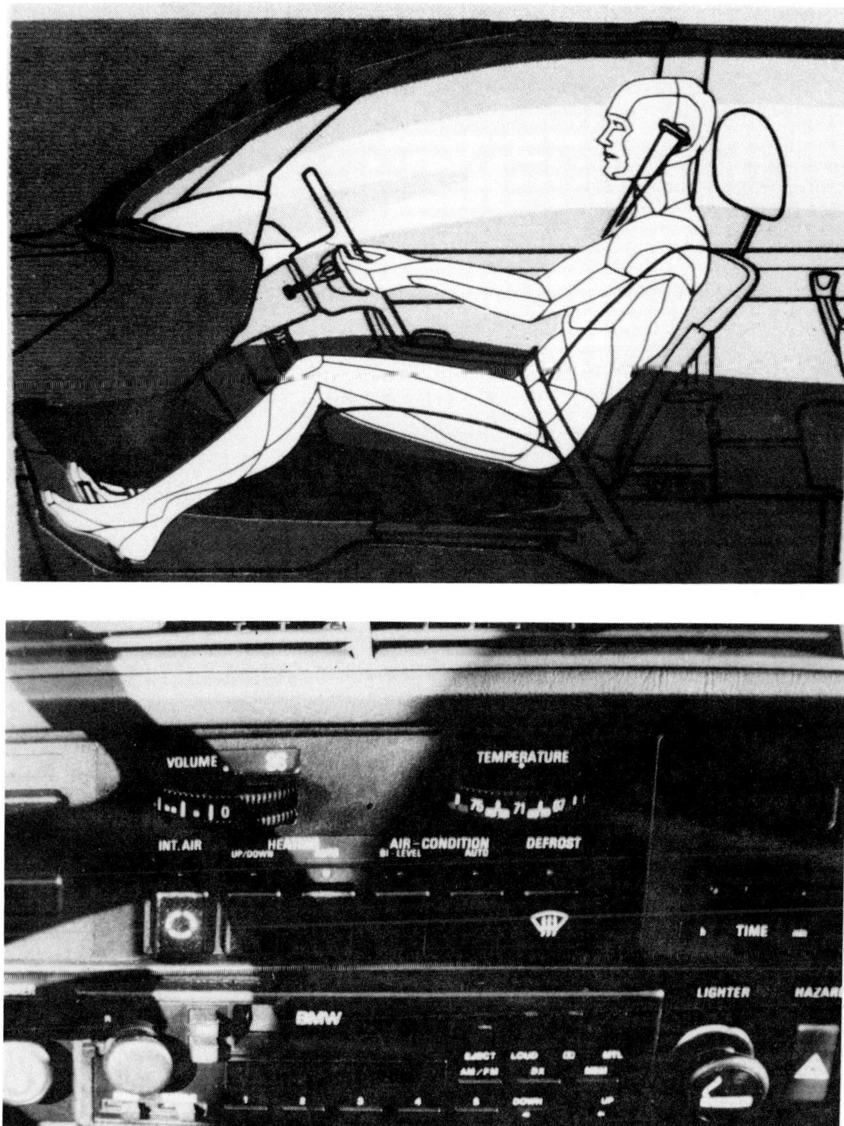

Figure 36. Schematic diagram and arrangement of controls for a car ventilation system. (Reproduced by kind permission of BMW AG.)

soft music) were drowned out by the racket assailing us from all directions. To help in giving us a relaxed feeling, the seat belts didn't fasten.

The worst was yet to come, though. The engines cut out three times as we were taxiing to the runway, and we were followed by a truck which had to start them on each occasion. By the time we'd taken off, the stewardesses had issued each passenger with two bottles of wine (one white and one red), and the anxiety level was receding rapidly. The actual flight lasted only forty-five minutes, but I hope never to have to experience such an agonising time again in my whole life. By the time we landed, I was a nervous wreck (despite the wine!). Since then, I've scrupulously avoided any journeys which might conceivably include a flight by that particular airline and have a standing instruction to that effect deposited with my travel agents!

Stress in a ship (unless you're in a naval war battle!) usually isn't a problem, though I've had the experience of seeing a few panic-stricken cruise passengers during gale conditions in mid-Atlantic. Ocean voyages in general, though, are expressly designed to eliminate stress. You sit there, day after day, with nothing to do except idly wonder what you'll eat for dinner or whether you should visit the casino that evening. A sea voyage is the best way I know to provide the relaxation that is so often missing from our modern frantic pace of life. It can, though, induce a different type of stress, the stress of boredom. If you've nothing whatever to do, there is a mental concern that you're wasting your life as you drift helplessly across the interminable waves, and psychological comfort diminishes in consequence.

The cruise lines reduce this problem by filling your schedule with an endless variety of activities, none of which matters in the slightest, so that you can feel fully occupied. The day is spent in such vital tasks as controlling the movement of a deck quoit or table tennis ball, holding a deck chair down, betting on the distance travelled during the day's run, checking the temperature of the swimming pool, participating in a scavenger hunt, listening to music, dancing, watching a film, playing bingo, entering a fancy dress competition, attending the passenger talent show, going to a church service, and inventing things to say to the same people day after day. It's a wonder there is any time left to eat with all that frantic activity!

On a train, there is the same kind of suspended feeling, especially if the journey is a long one. I find train travel very enjoyable, but only because I take some work along with me and only if the journey doesn't last too long. On one memorable occasion, I traveled from Toronto to Vancouver by train to have an opportunity to see the Rocky Mountains at close quarters and thoroughly enjoyed the trip. I did, however, have a private bedroom in which to sleep, and I had a stopover half way to pay a visit to a university where I had to give a lecture. Without those aids to my sanity, I doubt if the experience would have been half as pleasant!

One aspect of train travel which used to give me immense pleasure (and which has, alas, disappeared in Eastern Canada) was the chance to potter along to the dining car for a meal, served in the style of a first-class restaurant. The practice still exists in Europe (is it because life is so much more civilised, or is it just that they're not as quick to spot a "better" way of doing things?) but has unfortunately been sacrificed to the gods of efficiency, speed, and economy which are so universally worshipped in the parts of North America where I travel. I suspect it won't be long before the European trains suffer a similar fate; the signs of decay are there to see.

The subject of stress and its prevention wouldn't be complete without some mention of the car, that all-pervasive means of conveyance so essential to modern life. Despite all attempts to improve the situation, the act of steering a car along a busy street is still one of the most stress-producing occupations known to mankind. No matter how good your own skills, there are risks involved in driving as the road accident statistics clearly demonstrate. The usual problem concerns the other driver, who may be drunk, careless, asleep, inattentive, or at the wheel of an unsafe vehicle. (Though, naturally, no reader of this book would ever be guilty of such heinous crimes!)

Then there are the times when roads are icy, or slippery for other reasons, so that driving itself becomes more difficult. Road hazards, such as a sharp object that punctures a tire or a lump of wood that falls off the truck ahead of you, can occur in the most unexpected situations. I was once driving peacefully along a deserted road in the country when, ahead of me, I spotted an approaching light truck with something flapping in the back well. As it reached me, the object (which turned out to be an enormous sheet of steel) broke loose from its anchorage, flew up in the air, and headed straight for my face. I thought my last moments had arrived. Fortunately for me (and for you, on the incredible assumption that you're actually enjoying reading about it), the edge of the sheet snagged on a chain hanging providentially over the side of the tailboard, veered away, and hurtled off in another direction. The driver, unaware of his loss, continued serenely on his way, leaving me in a state of nervous delirium as I shakingly brought my own vehicle to a halt.

Now, how do we cope with that kind of stress? If you read the advertising matter for new cars carefully, you'll find the manufacturers concentrate on several comfort aspects, both physical and psychological. The design aspects of seating, controls, visibility in all directions, and other factors intended to make driving a pleasant experience are described in detail. Temperature control, leg room, height, reclining seats, and similar comfort-related features are specified. For mental comfort, style, colour, appearance, quality, luxury, and other attributes intended to create envy in an observer are subtly incorporated in the advertisements.

Then, for the faint-hearted passengers, the safety features which will give them an ability to relax in the vehicle are tactfully mentioned. Brakes, steering, road-handling capabilities, acceleration, durability, ease of maintenance, and such advantages are all given to reassure the prospective buyer that the car he is about to purchase can more than meet his needs. Even the most cursory examination will demonstrate just how much importance the manufacturer attaches to his customer's physical and mental comfort.

This matter of design leads us to consider another facet of transportation comfort: that of the finer things of life which we generally group under the single heading of aesthetics. When you travel, you want to feel that your surroundings are as pleasant as they can possibly be. You want to be surrounded by a decor that is not distressing to the eye or ear. Unfortunately, as we've already mentioned, no two people have the same opinion of what constitutes a comfortable environment, and this is true at the mental as well as the physical level. The task of decorating a public transport vehicle is thus not an easy one. What is needed is some neutral combination of colours which won't provide visual offence to anybody and yet won't seem so drab that the soul of the observer is subtly depressed by the surroundings.

Colours should blend harmoniously, not clash with each other. Contours of the interior of the vehicle should be interesting without being obtrusive (and, for physical safety, shouldn't protrude dangerously or be formed in sharp edges). The topic is so vast that there is not the opportunity in a general book of this kind to deal with it in any but the briefest detail, so the reader is referred to more specialised texts, such as those quoted in the reference bibliography at the end of this book, for further information. It's interesting to note in passing, though, that aesthetic design aspects occupy a significantly high proportion of the budget of every manufacturer of all kinds of vehicles, no matter what mode of transportation is being considered.

Finally, we come to the topic of efficiency. By this, I mean the ability to persuade the traveller that he is being moved to his destination in the most rapid manner possible, given the method of travel he has chosen. Speaking from a personal viewpoint, I dislike intensely any delay on a journey. If I'm stuck in traffic, whether in a car, a bus, a train, an airport, or anywhere else where delays inevitably occur, I get impatient and thus uncomfortable. It doesn't matter what the reason for the delay may be. I've been known, on occasion, to take a lengthy detour by an alternative road rather than sit waiting in a line of traffic. Even though the route I choose may in fact be much longer than the original one, I have the irrational feeling that, if I'm actually moving, my journey is progressing satisfactorily. I know it takes me longer, in many cases, to detour than it would to wait. I realise I'm being illogical, but it's no use arguing the fact. If I'm in motion, I'm travelling; if I'm stopped, I'm being held up.

How do the experts who move us around the country or the world deal with that kind of behaviour? Their first step should be to establish clear timetables and stick to them. This requires a careful calculation of all the arrivals and departures which could interfere with each other, followed by a plan to stagger them so that interference is non-existent or minimised. Then, unless it's absolutely impossible for any reason, departures and arrivals must always take place exactly as advertised, so the passengers feel they're not being treated with disdain. In years of travelling, I've never been so impressed with any mode of transportation as much as with the trains in France, Germany, and a handful of other European countries. On the many occasions when I've used them, they've never been more than two minutes away from the scheduled time of arrival or departure, even in the worst weather conditions. I wish I could say the same thing about our local services and about others I've used in various parts of the world!

The efficiency of moving people around should also extend to their personal effects. It's no joke to find yourself stranded in a foreign country with all your suitcases in some other place. It's happened to me once, and the experience isn't one I'd care to have to deal with very often. Airlines move millions of passengers and their baggage every year, and the fact that there are so few instances of loss is a good indication that the system is working efficiently. As must obviously be the case, a passenger who has faith in his ultimate arrival at the right destination, complete with all his belongings, will be far more relaxed (and therefore comfortable) than one who is worried throughout the journey about the possibility of an accident occurring, either to himself or to his possessions.

These, then, are the factors which should be considered when you want to improve the comfort aspects of transportation. A comfortable passenger is a happy one, one who'll return to travel with that particular organisation again, and the experts are fully aware of this simple fact. That is why there is such an insistence on passenger comfort in the design, service and operating arrangements of all vehicles for moving people by land, air or water. In the same way, you should be aware of the ways in which your own comfort can be maximised. A comfortable driver is a safe driver, and you may well find that attention to comfort could literally save your life. In yet another way, comfort and survival may be virtually synonymous terms!

CHAPTER 12

TEXTILE FACTORS IN COMFORT

IT SHOULD BE obvious by this time that, for just about every aspect of comfort, textile products are an essential component which can't be omitted from any consideration of the subject. We've already examined briefly the fashion aspects of clothing, and it's now time to take a closer look at the textile properties which are incorporated in these and many other goods that contribute to human comfort.

Textile materials are used widely in the home, in industry and in transportation, as well as merely in clothing, and their properties are a vital part of the reason for their widespread use. Everywhere you go, you'll continually come across them. They're so intimately bound up with human existence that it's impossible to envisage modern life, in any of its multifaceted areas, without an impressively large use of textile products. Look around you, wherever you are, and see if you can find any situation on earth where no textile object is located if human existence (and hence human comfort) is to find a place. The materials which pervade so much of our life depend on the behaviour of the textile components for their usefulness, and their properties can be classified into those of fibre, yarn, or fabric for ease of examination.

Fibres are the basic units from which all textiles are produced, several different varieties being shown in Figure 37. They're tiny structures, ranging in length from about a centimetre to about 40 centimetres for most natural fibres, though silk filaments have a length of several kilometres and all synthetic filaments can be produced in an infinite length. In cross-section, they vary from perhaps 20 to 50 micrometres in diameter, a human hair (for comparison) being about twice as coarse as the thickest textile filament normally used. As you can see in Figure 37, wool fibres have a scale surface structure, cotton fibres have a twisted-ribbon appearance, and synthetic filaments are, in general, uniform cylinders.

These units are gathered together in long bundles, with twist to keep the structure stable in the case of fibres of finite length, by a complex series of operations which results in a yarn as the end product. A schematic diagram of the simplest type of yarn is shown in Figure 38, in which photographs of a variety of more complex yarns are also shown. In each case, the spiral arrangement of the individual fibres around the major axis of the yarn is clearly visible. Yarns are then woven, knitted, or interlaced by some other means, to form the fabric used in manufacturing the textile products so valuable to our continuing existence. A fabric, as may be seen in Figure 39, is a porous structure of complex form with holes and smaller interstices distributed throughout the fibrous lattice.

All the myriad products are made in a similar way, with variations for the specific end uses if necessary. In clothing, fabrics are worn from skin to outermost coat, and a range of fabric specimens, typical of those used in a winter clothing assembly, is shown as Figure 40. The materials are arranged so that the progression from skin to outer surface is clockwise from the top, several types of fabric being shown at each stage. Knitted underwear, woven shirt or blouse, knitted or woven suits, skirts or trousers, woven or felted linings, woven or knitted coatings and water resistant outerwear are represented, and some or all of such fabrics are almost invariably worn in most countries of the world where climate makes the use of warm clothing necessary.

In the home, tufted or woven carpets and woven or stitch-knitted curtains, upholstery or bed quilts abound. Woven sheets, table linens, and towels for hands or dishes make life more pleasant for us, either physically or psychologically. In transportation, the carpets, wall coverings, upholstery, uniforms, noise- or heat-insulation materials, engine filters, tire cords, and other vital elements of construction make life more comfortable on our voyages.

Industry depends on the same kind of auxiliary components for

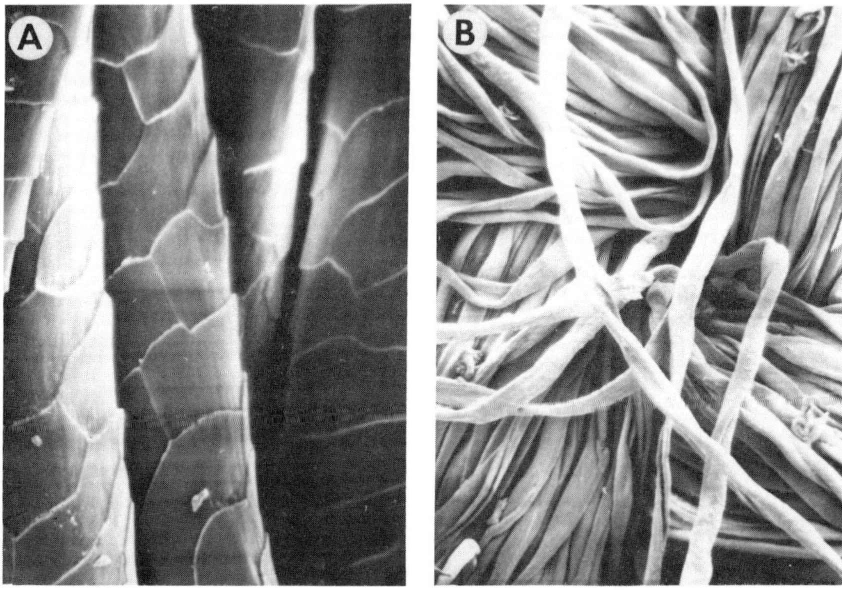

Figure 37. Photomicrographs of several textile fibres: (A) wool, (B) cotton, (C) acetate rayon, (D) nylon, (E) polyester, (F) cuprammomium rayon, and (G) Antron nylon. (The photographs of wool, cotton and polyester are reproduced by kind permission of Doctor Anne Wilcock.)

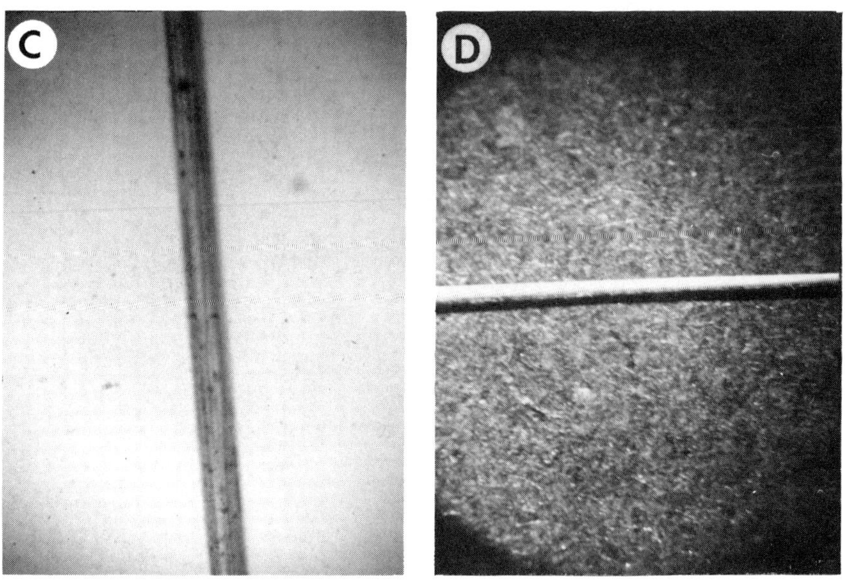

E

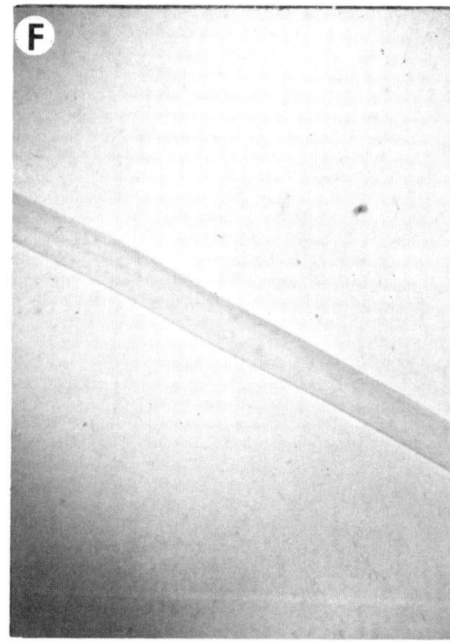
F

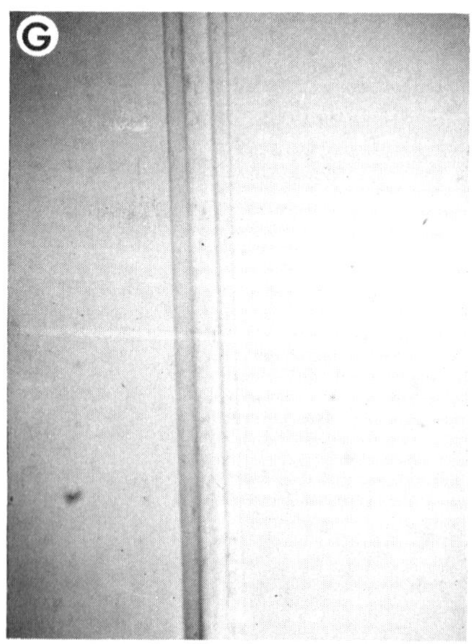
G

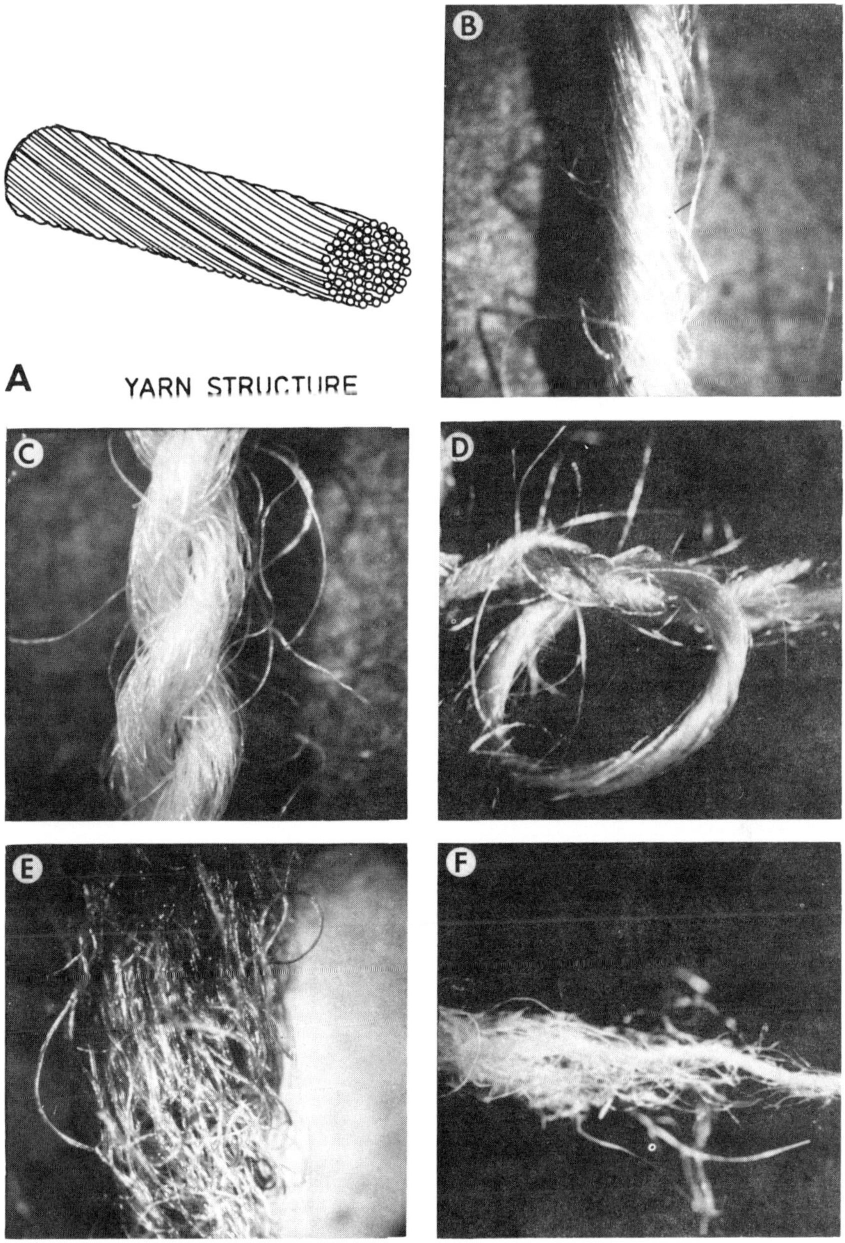

Figure 38. Yarn structure showing spiralling fibres: (A) schematic diagram, (B) singles yarn, (C) 2-ply yarn, (D) loop yarn, (E) flock yarn, and (F) corespun yarn.

Figure 39. Characteristic fabric structure.

its successful continuation, and relies on textile products to supply them. Industrial textiles make up a significant part of the manufactures of the textile mills. Ropes, tarpaulins, conveyor belts, paper-making felts, and filter cloths are all familiar enough to need no description. It's no idle boast on the part of the textile industry that it must, quite literally, provide the world with virtually all its needs. Even the food reaching our table couldn't do so if there were no tires (which contain textile materials in the form of tire cords) for tractors or trucks, fabrics for packaging during transport, clothing for agricultural workers, nets for fishing, and all the miscellaneous hidden textile products in any piece of machinery. Truly, our very existence, let alone our comfort, is vitally affected by these ubiquitous materials.

First we should examine the properties which are related to the fibres themselves. These can be subdivided, in turn, into chemical and physical ones. The chemical ones of interest include absorbency and reactivity, both of which can exert considerable influence on comfort in several ways. Absorbency increases the physiological usefulness of a garment, in that it helps the process of removing perspiration moisture and hence maintaining a stable body temperature by thermoregulatory cooling. It also influences the ease of dyeing, so enabling the manufacturer to provide a pleasing range of colours to increase our choice.

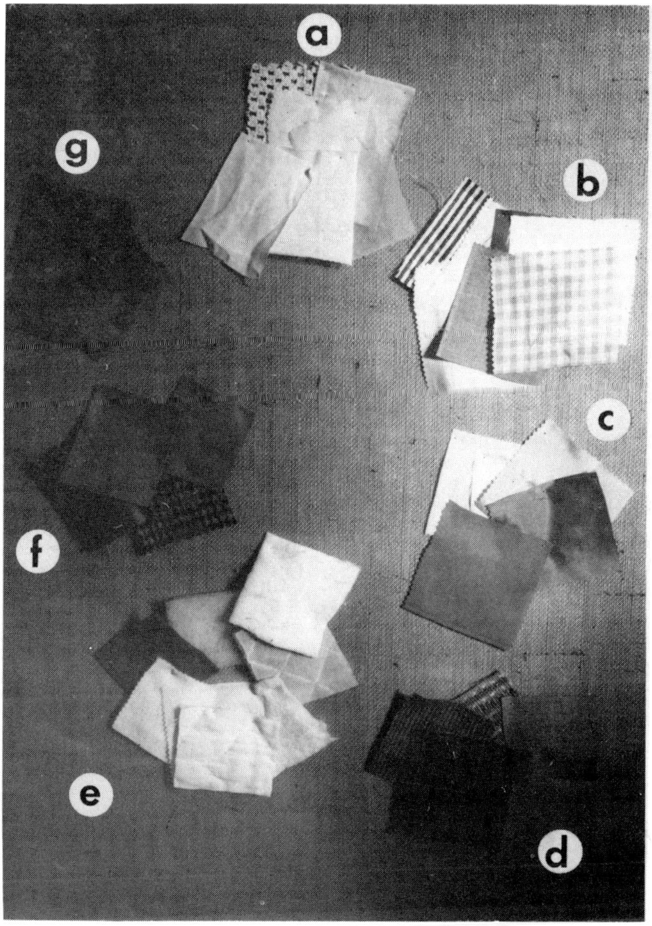

Figure 40. Fabrics useful for winter clothing assemblies (clockwise from the top): (a) underwear, (b) shirtings, (c) jacket linings, (d) suitings, (e) interlinings, thermal linings and paddings, (f) coatings, and (g) outer water-resistant fabrics.

The finishing of textile goods, which gives them such desirable properties as ease of maintenance, flame resistance, anti-static nature or similar ones, is also strongly dependent on absorbency in most cases. Chemical reagents are applied to the fabric and, when they've become soaked into the fibre pores, are attached more firmly (and thus more permanently), either by a chemical reaction between the finishing compound and the fibre, or by heat which brings about a polymerisation, or by subsequent additional chemical treatment with a second reagent.

Reactivity determines how useful a material will be as a protective covering, for example, against chemical or microbiological hazards and heat. It also governs the ease of removal of stains, the launderability, the ability to withstand bleaching agents or solvents, and the likelihood of durability or appearance being affected by atmospheric and environmental conditions.

This chemical behaviour depends entirely on the structure of the molecules from which the fibres are made, and thus on the fibre content of the article. Each molecule in the structure is a building block, shaped like a long, thin thread, which can attach itself to other molecules, either end to end or sideways. The molecules contain side-chains, additional molecular units sticking out (as you might expect from their name) in a sideways direction. On these chains are the reactive sites which govern the chemical behaviour of the fibre, and two effects combine to give each fibre type its characteristic properties.

The first of these is the spatial arrangement of the atoms. Figure 41 shows the structure of two types of fibre (wool and polyester), and it's easy to see that the wool, with its spiral arrangement and a high proportion of side-chains (and hence many pores within the structure), is much more bulky than the compact polyester with fewer side-chains.

If the side-chains are bulky, they'll prevent the molecules from packing tightly together, as shown in Figure 42. There are then very few chemical bonds between adjacent molecules, so there isn't much holding the structure together and the fibre tends to be a weak one. On the other hand, there is ample space for water, dyestuff, finishing agent, or chemically reactive substances to find a way into the structure, so making the fibre absorbency and reactivity likely to be higher.

At this point, the second factor comes into effect. If the reactive sites on the side-chains are chemically such that water, a dye molecule, a finish, or a reagent can enter into a reaction easily, then the fibre will tend to unite chemically with the penetrating substance and be absorbent to it. A common process, the permanent-press finishing of a cotton fabric, depends on the ability of a chemical reac-

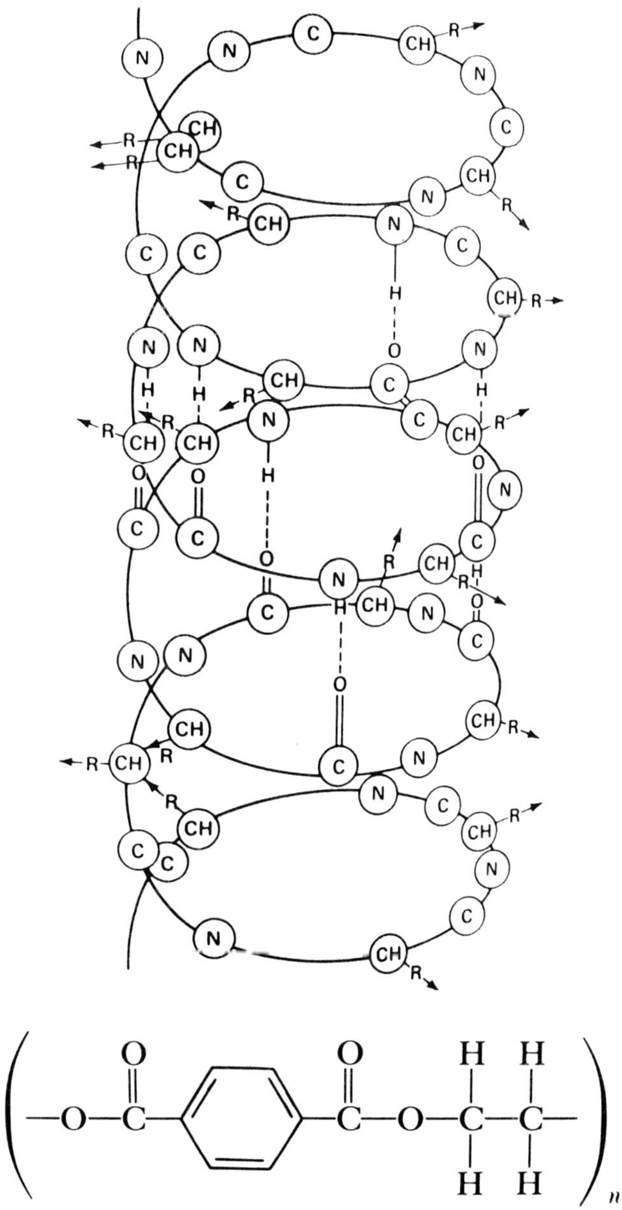

$n = 80\text{–}100$

Figure 41. Chemical structures of wool and polyester.

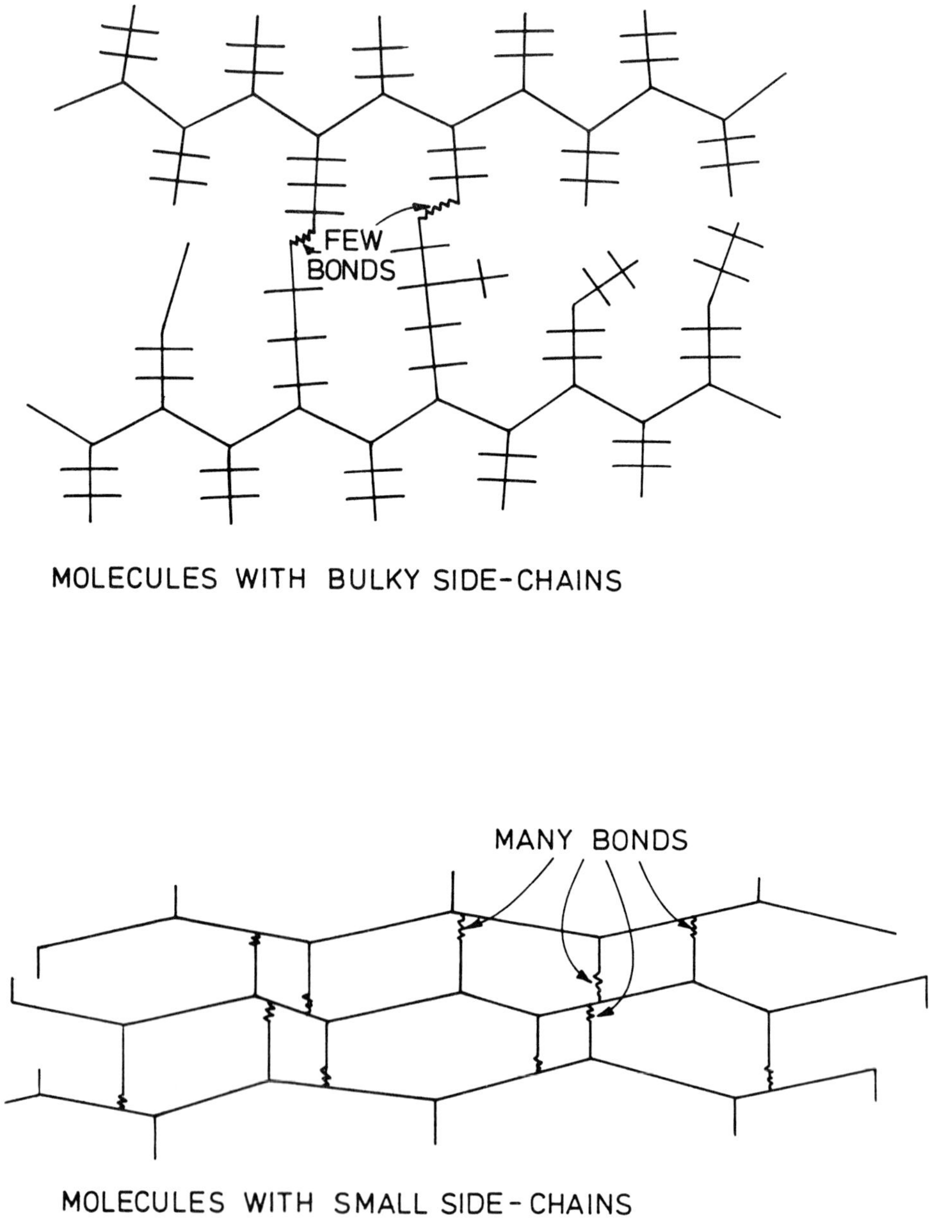

Figure 42. Molecular packing with bulky side-chains.

tion to destroy molecular bonds in the cotton cellulose and reform them in new positions, so that the cloth tends to be more stable in a pressed form than in any other one. The part played by the side-chains in permitting access of reactive molecules and in providing a site for reaction to occur is a vital one. Thus, the chemical structure of a fibre is very influential in determining many of its comfort-related aspects.

One important property which belongs somewhere between chemical and physical in behaviour should be considered here. If a fabric contains many tiny spaces, then surface tension effects (as described in Chapter 5) become very important. Even if the fibres are unable to absorb large quantities of water, the liquid will penetrate through the structure by capillary attraction. The process is known as *wicking* and can compensate considerably for poor absorption in a garment. Consider, for example, the case of a shirt or blouse worn on a hot day. At regions of high perspiration (under the arms, for instance), a cotton fabric will absorb moisture and transport it to other areas where it can evaporate from the cloth surface.

A normal polyester garment in the same location, however, would soon become saturated, because the water produced isn't absorbed and thus remains in the vicinity, unable to escape at the low rate of evaporation possible in the humid atmosphere. The two situations are shown diagrammatically in Figure 43, where the differences in behaviour are clearly seen. If, however, the polyester filament is textured (that is, given an artificial crimp or kinked character, as shown in Figure 44), packing of the individual filaments will be less compact and many pores will exist in the fabric structure. Wicking can occur via these pores, so liquid can be transported, once again, away from the region of high moisture concentration and given a better opportunity to evaporate. The wearer will then be more comfortable, since the onset of tangible wetness will be postponed by the dissipation of moisture.

The physical properties, too, are important. We've seen how fibre strength (critical in such matters as durability) depends on chemical structure, but it's also influenced by fibre fineness. If a yarn is made up from many fine fibres, rather than fewer coarse ones, there will be more frictional contact between adjacent units and the cohesion, which directly affects strength, will be improved.

In addition, fibre fineness influences flexibility. A fine fibre will bend far more easily than a coarse one and so will the yarn made from it. In consequence, the softness, drape, flexibility and general feel (or hand, as it's called in the trade) of the fabric will change, with a noticeable effect on both physical and psychological comfort.

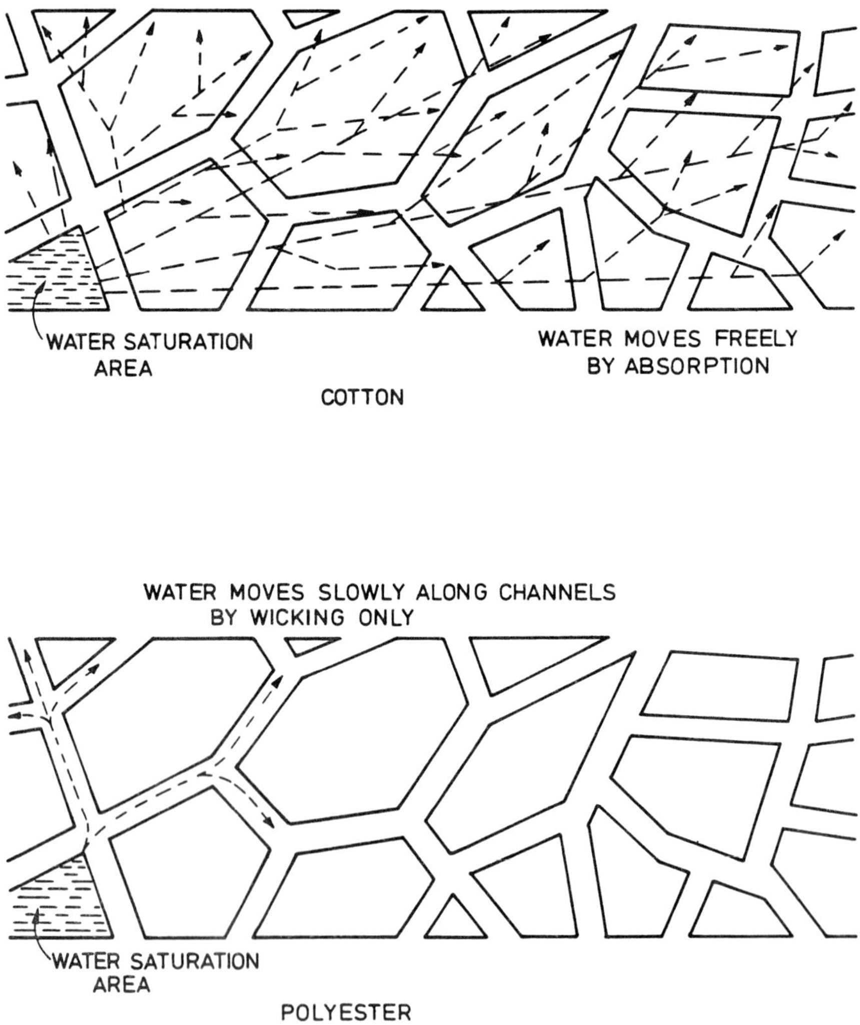

Figure 43. Wicking in cotton and polyester.

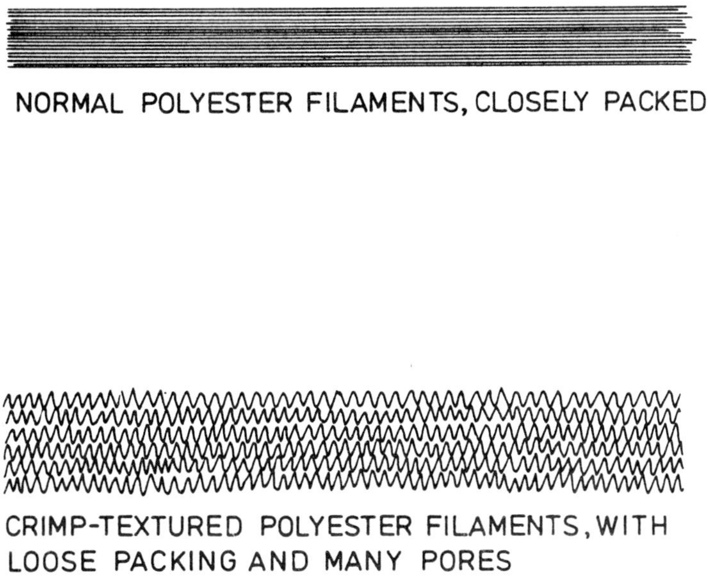

Figure 44. Crimp in fibres.

These properties, apparently so simple, are in fact highly complex. The drape, or the way in which a fabric hangs freely in attractive folds under the influence of gravity, is a function of its mechanical structure. The buckling that takes place in the plane cloth surface, as a result of the complex forces existing between structural elements, is an intricate three-dimensional deformation and very difficult to describe mathematically. Flexibility, the way in which a fabric folds when a force is applied across a plane, is somewhat simpler to characterise but not yet completely defined. The most difficult property of the three, however, is softness, depending as it does on the subjective element, with all the problems we noted in Chapter 6. The electrical, chemical and other reactions produced within your brain as your skin moves across the surface of a piece of cloth and its undulations provoke such a multitude of neural responses that it's impossible to attempt to analyse or fully comprehend them.

Thus, no quantitative definition can be given, or even remotely envisaged, for the property of hand. All that can be done is to make a few general statements about specific fabrics in comparison with other ones. A fine fibre, in general, will impart a softer hand to the garment made from it, but the cloth will also tend to crush more easily, thus being likely to cause increased psychological discomfort as a result of flawed appearance. Fibre crimp, the way in which the individual fibres zigzag along their length, influences the ability to move moisture (as discussed earlier in this chapter) or trap heat and thus the thermal comfort. A fibre with high crimp, like wool, is an excellent thermal insulator and allows moisture to move more rapidly through the structure; so it is more comfortable than, say, a synthetic with no built-in crimp. Many synthetic fibres are deliberately treated after manufacture, as we've seen, to give them an artificial crimp to improve their comfort behaviour, and such products are roughly intermediate in properties between a wool and an untreated synthetic fibre. Figure 45 shows a series of photomicrographs of fabrics made from each of the three types of yarn, and the difference in porosity can easily be envisaged.

The properties of the yarns must also be considered. The relevant ones are fineness (the cross-sectional dimension) and twist (the number of turns per unit length given to the fibres in making the yarn), both of them affecting mainly the softness, durability and appearance characteristics. A selection of coarse and fine yarns with loose and tight twist has already been shown in Figure 38, and a glance at that figure will serve to indicate the variety available. It's possible to make general statements about the effect of yarn size on fabric behaviour. A finer yarn will bend more easily than a coarser one so that the fabric made from it will feel softer, as for the fibre effect. On the other hand, it will be weaker (other factors being constant), so it may not last as long. A yarn with high twist, conversely, will last longer because it will stand up better to abrasive wear but will feel harsher to the touch. The actual end-use requirements will determine, in both cases, which trend is preferred.

Finally, the bulk properties of the fabric should be included, since they're probably more critical in determining comfort than those of either yarn or fibre alone. In Figure 46, a fabric with a compact structure is shown with a "lofty," or loosely structured, one to illustrate the different appearance of the two. Other properties beside appearance, though, are clearly important in considering the com-

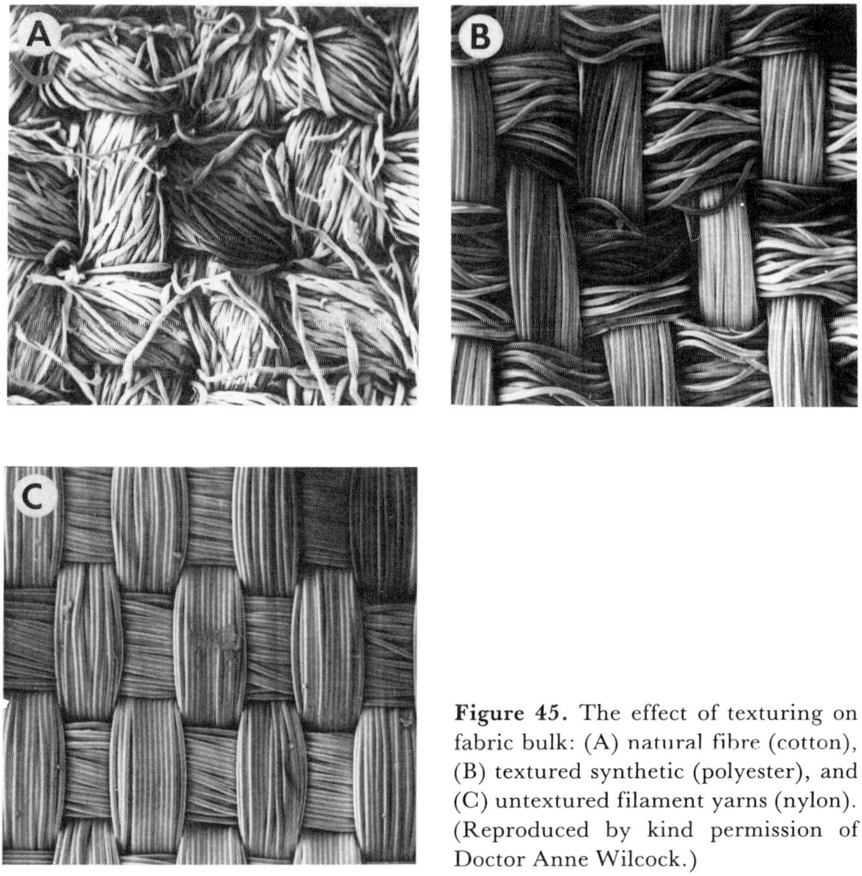

Figure 45. The effect of texturing on fabric bulk: (A) natural fibre (cotton), (B) textured synthetic (polyester), and (C) untextured filament yarns (nylon). (Reproduced by kind permission of Doctor Anne Wilcock.)

fort behaviour of the fabrics. The most influential ones are the weight, thickness, yarn spacing, method of construction, subsequent treatment, and dimensional stability, and I propose to discuss briefly each of these in turn.

Weight and thickness usually change in proportion to each other and so may be considered together. They increase thermal-insulation ability (since an increase provides additional air pores to trap body heat), but they add an extra load for the wearer to carry around and so tend to increase fatigue and thus reduce physical comfort. Weight is governed not only by the amount of material used in making a fabric but also by the density of the material. A fi-

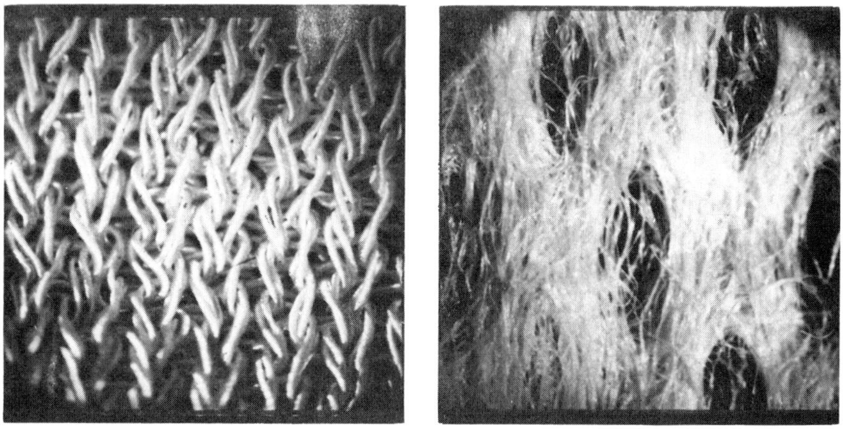

Figure 46. Fabrics with compact and lofty structures compared.

breglass cloth, for instance, would have about three times the weight of a cloth of identical construction made from polyester. The optimum weight of a garment will be one in which adequate insulation is provided without placing undue loading on the strength of the wearer.

Yarn spacing also exerts some effect on thermal insulation. Figure 47 shows two wool cloth samples, placed side by side, with a light shining through part of each one. The two look similar in their weave fineness at first glance but are seen to be very different in appearance when a light is shone through them. Just as the light passes through them in different amounts, so also do air, water vapour and heat. Thus, a very open structure will tend to allow heat to pass through easily, so making the wearer cold in winter and too hot on a warm summer day. On the other hand, a close structure will tend to prevent the escape of moisture, so may reduce perspiration comfort. Additionally, the actual weather conditions, particularly wind speed, will have a drastic effect on the insulating ability of a fabric as a result of the wind-chill factor discussed earlier.

Modesty, too, may be a reason for selecting a particular yarn spacing. If a flimsy, see-through effect is desired, then an open structure is ideal. On the other hand, if nothing is worn underneath the fabric, the effect may not be exactly the right note to aim for in a job interview at the local monastery. Spacing can also cause a change in

Figure 47. Yarn spacing in high- and low-quality wool cloth.

durability to arise. A tightly woven fabric will tend to be stronger than a loose one, mainly as a result of the decreased likelihood of snagging of yarns as they rub against any nearby surface.

The usual term used in the industry to describe the overall appearance of a fabric's surface, in terms of the proportions of solid material to air space, is *cover*. It influences most of the physical comfort properties, such as water flow, air permeability, wind chill or insulation, and some of the psychological ones, like modesty, as well. A fabric with high cover (examined under a microscope) has a very compact structure, but one in which cover is low appears to be full of holes. If you examine large-scale photographs or microscope images of two sheeting fabrics, as illustrated in Figure 48, the difference between the high quality percale (made from finer yarns and woven more tightly) and the muslin of lower standard (and hence lower expectancy in terms of the service likely to be given) is clearly visible.

The fabric construction method is another variable which can modify durability, thermal insulation or other comfort-related factors. A woven material is generally tighter than a knitted one and a felted one less open than either. The newer techniques, such as stitch-knitting, laminating, or bonding, each have their own individual characteristics. A complete discussion of all the different types of construction method (and there are some fifty or more of them!) is, unfortunately, outside the scope of this book but is available in

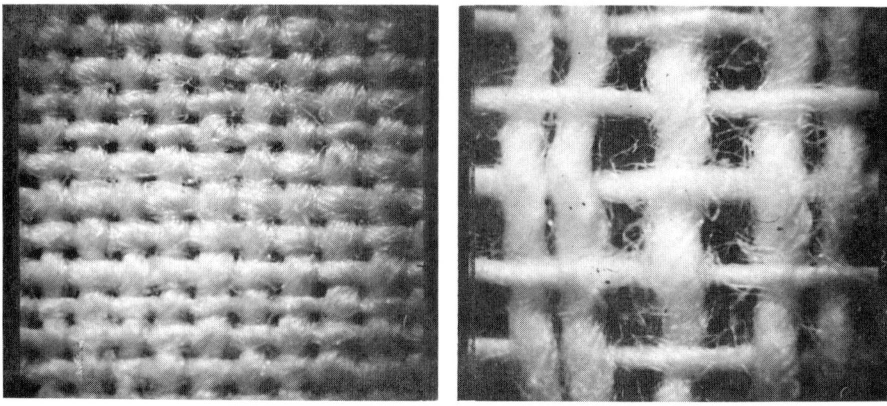

Figure 48. High- and low-quality sheeting fabrics.

any standard work on textiles. In general, though, a fabric which is open in structure will be a better insulator in still air than a closed-construction one, but the reverse is true if any air movement (including the movement taking place as you walk through still air) is present. The reason is simple. The air cells that hold the heat are larger if the fabric is open, but the protective still air layer is swept away more easily in such a fabric.

In considering durability, it's important to remember that the loops of a knitted structure or a stitch-bonded one are likely to snag. Thus, when the fibres are dragged from within the structure of the cloth to the surface by frictional contact, the action will tend to create unsightly faults which eventually cause the material to be rejected because it no longer looks good enough to wear. A woven, laminated or bonded one, on the other hand, won't tend to snag easily and so will be visually attractive for a longer time. A felted fabric, with virtually no open pores but lots of tiny, enclosed ones, will provide excellent insulation, but the fact that it's only loosely held together by inter-fibre frictional forces means that it will tend to separate and stretch out of shape easily.

Even the type of pattern used within a construction method can have an appreciable effect. A satin weave, with long yarns "floating" over the surface, will snag more easily than a plain or twill, where the yarns are all tucked securely inside the structure, and so will tend to become unattractive more quickly. It will, though, have a more luxurious appearance and hand when it's first put into service and will give more aesthetic satisfaction to the wearer than will fabrics in either of the other weaves. Figure 49 shows the differences in appearance between these three common fabric types. As can readily be envisaged, the tighter structures of twill and plain weaves impart stability, and the twill, in addition, has a characteristic diagonal ridge feature that gives more interest to its surface. Twill cloths are used in garments where strength is important and surface texture is needed to add interest.

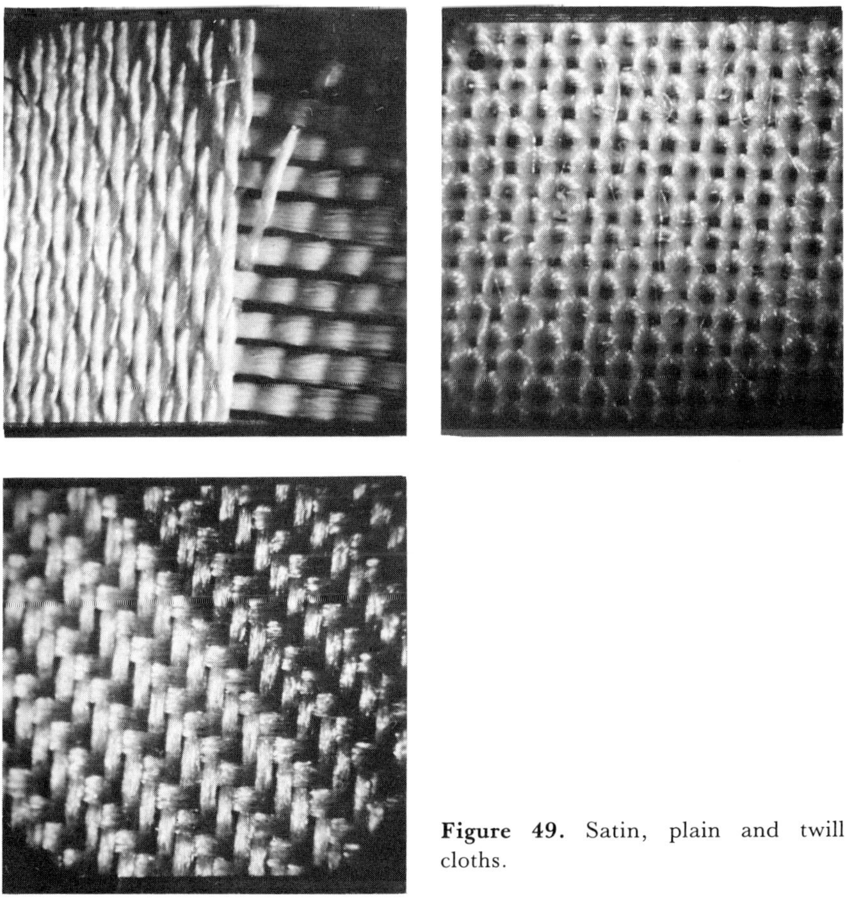

Figure 49. Satin, plain and twill cloths.

The texture of a fabric, governed too by construction parameters, will exert some influence on comfort. A cloth that is irregular on the surface will be far more interesting to look at than a plain one. A penalty is paid, though, if it's worn next to the skin and the protruding bits of fabric stick into the protruding bits of body, usually with the most undesirable visual and tactile effects. Texture can, if properly chosen, increase the psychological comfort by imparting a slightly rough hand to the garments, and it will certainly assist in reducing the shine or transparency of a fabric. It's particularly effective in giving visual attractiveness to a furnishing scheme by providing interesting variations in a room. As usual, it's a matter of being knowledgeable in the fabric chosen and the way you use it.

After the cloth is made, treatments which can have some effect on comfort are frequently applied to it. It can be impregnated with various chemical compounds to impart easy-care, flame-resistant, anti-static, water-resistant, or other properties of a desirable kind. It can be dyed or printed, embossed or given artificial texture, and brushed or sheared. Most of the benefits of such treatments are self-evident, and it's obvious from a glance at the catalogue of any supplier of textile auxiliaries that finishing is both a popular and a necessary process within the industry. There are literally thousands of compounds on the market which are used regularly by textile manufacturers.

The finishes intended to make the fabric safer or easier to look after will increase mental comfort by reducing hazard or workload. Those providing protection against the elements or natural phenomena will make the wearer more comfortable physically. Brushing improves appearance and thermal insulation. Shearing makes the cloth look neater. Design and colour aid psychological comfort. In every case, the manufacturer is trying to make his product more appealing by persuading the buyer that it has some inherent property which is desirable because it will give comfort of some kind to the owner of the fabric.

We come now to one aspect of fabric production which shouldn't be overlooked: that of dimensional stability. A fabric, when manufactured, can be subjected to many physical stresses, all of which leave tensions within the structure. If these tensions aren't removed during the final stages of processing, they remain dormant within the fabric. As it's used and allowed to relax (when it becomes wet in the rain or during laundering, for example), the tensions can be released slowly. When that happens, the structural integrity is disturbed and the material deforms out of shape.

Incorrect finishing of a fabric, followed by subsequent laundering under supposedly correct conditions, can cause it to shrink and become distorted out of its true shape. A garment made from it may shrink, or stretch, or may undergo a combination of both processes at different places so that it resembles something out of a secondhand clothing store taken straight from the pages of Dickens. The usual way to prevent this problem is to apply a pre-shrinkage treatment to the fabric before making it up into a garment, but this isn't an infallible solution. If the stresses aren't completely relaxed by it (as is often the case), trouble with dimensional stability can still arise.

The design imparted to a fabric can influence comfort, usually in a psychological sense. The cloth may be dyed, printed, or treated with a finish to give it the changed appearance intended. Alternatively, the method of interlacement of yarns can accomplish the same aim. Affected by the combination of dyestuff and fibre content chosen are not only the colour and texture but also the long-term satisfactory performance of the material. If the dye fades or crocks (that is, rubs off when abrasion takes place), the article will tend to be rejected earlier because its appearance makes it unacceptable for wearing in public. A garment which has undergone fading in areas as a result of light exposure is often rejected as no longer suitable for wear or use. A poorly designed structure, which can fray or pill, can bring about the same decision. Changes occurring as a result of wear (fading or disintegration when strength loss takes place as a result of exposure to light, for instance) can also influence the life of a garment and the mental comfort of the person wearing it. A cotton shirting or tenting, for example, can be degraded, by exposure to weather, to the point where it may disintegrate almost at a touch.

Any consideration of textile factors affecting comfort would be incomplete without some reference to industrial textiles. We take these even more for granted than we do our household ones, but we couldn't remain comfortable (or even alive, in some cases) without them. The tire cords on which we drive are literally holding death at bay. The conveyor belts that move escalators or luggage at the airport save us the effort of climbing stairs or carrying heavy cases, an effort which could be fatal to somebody with heart disease. The threads that hold garments, upholstery or surgical wounds together are textile products; so are the ropes on which a mountaineer, a camper, or a seaman depends for the successful pursuit of a hobby or a job. The filter cloths keeping the air inside an industrial plant free from dust, or isolating a human being from a virulent infection, couldn't exist without the textile industry. Nor could the insulation in your home, your car, your clothing and anywhere else you need to keep cold or noise at bay. Truly, the industry has an important part to play in keeping us in comfort.

We've seen, then, a few of the problems which the textile manufacturer, or his engineer, has to deal with in the task of bringing his clothing, household, or industrial products to market. He is coping with the vagaries of fashion, that mysterious mass hysteria which drives us all to buy clothes we wouldn't have been seen dead in the year before, vagaries which may make his goods redundant at a minute's notice.

He's also forced to use raw materials which are difficult to handle because their units, the individual fibres, are so small. His machinery, costing perhaps two or three hundred thousand dollars per unit, is temperamental, highly specialised, and expensive to replace, maintain or run. He has to be ready to switch to a new line of goods, often totally different from what he was producing the previous week, using equipment not really designed for the purpose. He must risk large volume runs in order to meet the demand if his product is successful, but he is left with a vast quantity of useless stock if fickle consumers turn from his items. Under such circumstances, it's nearly a miracle that any manufacturer stays in business, let alone manages to make a profitable living. We're indeed fortunate to have a textile industry to help us in our lifelong search for increased comfort.

CHAPTER 13

COMFORT IN EVERYDAY LIFE

THE MORE observant or astute of my readers will have noticed that this book has thirteen chapters. This is no accident. It was done deliberately (or so the author claims with the wisdom of hindsight) to emphasise a very important point. I trust that the superstitious ones of the world will not be unduly alarmed (and, more to the point, will not refuse to buy the book!) at this ominous portent.

The important point, of course, is one which has been mentioned occasionally in the preceding pages. It is, quite simply, the fact that comfort can be diminished by the most unexpected and apparently trivial event. Taking the rather contrived example of the chapter numberings as a somewhat bizarre instance, we could embark on an imaginative journey into the realms of superstition, discussing all the strange fads which give people shivers of fear that can destroy their comfort. Numbers that must never be used in their presence, what to do if a black cat crosses your path or the salt is spilled, the dread of Friday the thirteenth, and so on, are common concerns in our society. It doesn't matter that there's no rational reason for the fear. The fact that it exists is enough to change the comfort level of somebody who would otherwise be perfectly at ease.

Now, from this example, we can broaden the scope of our enquiry to look at all the different manifestations of comfort that appear in our daily life. We must never forget, though, that there is such a tangled web of interconnection between all aspects of comfort that we can never hope to analyse any given situation completely. All we can do is to provide as much information as possible about what makes people comfortable or uncomfortable and then hope to recognise the reasons for our own comfort or discomfort at any particular time and maintain or rectify these conditions.

To carrry out such an ambitious task properly would, of course, need a second book. The topic is so vast that it's impossible to do justice to it in the space of one chapter. If you'd be kind enough to write to the editor, clamouring for further information, I'll wait stoically on the sidelines for my call to action when the level of demand has risen to an irresistible crescendo. In the meantime, my contract says I have to get on with this book, so I suppose I'd better finish the chapter.

In any discussion of comfort needs, as I've tried to point out several times already, there will inevitably be some duplications or compromises. The comfort needs are so complex that the same solution may work to satisfy more than one of them. Conversely, a solution for one type of comfort need may well destroy or reduce comfort of another kind. Take, for instance, the action of walking quickly to remain warm in winter. It enhances thermal comfort, but it may well make the walker breathless (and hence uncomfortable in the lung area) or give him aching leg muscles (and hence cause discomfort of yet another kind).

If the alternative solution of putting more clothes on is tried, those two particular comfort problems may not arise, but the garment may well cause an itch on the skin or a choking sensation around the neck. Thus, comfort problems tend to be difficult to solve because of such incompatibilities.

What I propose to do, then, is to take the subject matter of each of the first twelve chapters of the book more or less in sequence and discuss the ways in which comfort can be optimised in each of the different types of situations encountered. The result will inevitably be highly subjective, since it will consist largely of my own opinions and of solutions that I've found effective. Your own needs may differ greatly from mine, and the solutions I propose may not work for you. Nevertheless, I hope that the simple act of focusing on a problem, and on one possible solution, may give you the opportunity to think more deeply about the situation and may give you some idea of how to work towards a possible solution that will be effective for you.

I suggest we can all agree that mental comfort must be considered first. If you're not happy with the way you feel within yourself, you can't be truly at ease with the rest of the world, no matter how well you look after your material comfort. The secret of being comfortable in this way is to look at what you've achieved, or what you're striving for, and decide that, no matter how insignificant it may appear to be, it's worthwhile. You're a human being, and that in itself is a major triumph of biological evolution. You may be the most worthless human being on the planet, but you're still a human being, and that makes you a wonderful, almost miraculous, creation. Nobody can take that away from you as long as you're still alive.

This starting point can be the basis for developing your own self-confidence. If you can find the assurance to believe that you matter, that you're important, then confidence will follow. The simple fact is that every human being on this planet can be useful, or even necessary, to somebody else. Some of us have the power to sway the lives of millions, whether we are politicians, soldiers, or preachers. Others are important to a smaller number. Life would be less enjoyable for many without their favourite film star, pop singer, football hero, or other such object of adulation. A teacher can influence the lives of many students, a few at a time, and that influence can flow outwards to others in unseen ripples.

Even the smallest child can produce an effect out of all proportion to its apparent size. I recall, on one occasion, walking though Central Park in New York with my family (in the days when it was safe for families to walk in Central Park!). An elderly lady rushed up to us and insisted on giving me a silver dollar for my oldest daughter. I tried to refuse it tactfully, since she was obviously not wealthy, but she insisted. Apparently my daughter, aged about six, had smiled at her in a friendly way, and it was the first time that had happened to her since she'd arrived from Puerto Rico some weeks earlier.

So take heart. You have some value to the rest of the world. You may never realise it, but you're constantly influencing others and providing comfort for them in ways you wouldn't even recognise if you were aware of them. The blows which the world delivers to make us lose our confidence are only part of our existence. It's just the same as with any other aspect of comfort. The bad aspects (that is, the discomfort of a devastating attack on our confidence) are always felt more keenly than the good ones which give us a better feeling about ourselves.

Now, how do we acquire confidence? As with all other human traits, it's a matter of experience. We have to learn what we can do well and concentrate on doing it well. At the same time, we have to recognise our limitations and avoid doing those things which are beyond us. The process of growth involves finding more and more things that we can do well so that we're prepared to tackle more and more difficult problems as we progress along life's path. Confidence is a matter of keeping pace with our growing abilities. If we always stay close to our limitations and restrict our new steps to small ones that don't push us too far emotionally, mentally or physically, then we'll gain confidence in our own abilities. It's only when we're faced with a situation that is totally outside our capabilities that our confidence is shattered and we feel desperately uncomfortable or shy as a result. If we train ourselves to think rationally and face up to our inmost fears openly, then confidence will come.

There are many ways of helping to develop confidence. Some people seek the reassurance of religion. Others prefer to rely on the strengths of philosophy or on the abilities of a hypnotist. Yet others refuse to accept any external aid, preferring to cope with the problem of low confidence themselves. I can't decide which approach is best for you, but I can give you one piece of advice to help you make the choice. If you feel that your chosen method of developing confidence is making you unhappy, or if you experience doubts about its success, the chances are that it's not for you. In particular, if the aid doesn't work without some additional support, then forget it. If, for example, you can't face the world without a drink, or without spending hours agonising over what you should wear, or without hiding under a mask of makeup, then your confidence isn't as good as it might be. Try to develop it in some way that doesn't involve crutches of that kind if you really want to be mentally comfortable!

An additional aid to increased confidence is the development of breadth of mind. There are many ways to accomplish this. The simplest one is merely by reading, an activity in which you're engaged even now. Read widely to give you familiarity with new ideas, ideas which you'll be able to discuss with other people. One of the best ways to develop confidence is by talking to others, and if you've something worth talking about you'll be an interesting person and hence will reap the rewards of increased confidence when people seek you out.

Breadth of mind can also be increased by travelling, as many young people are finding out in greater and greater numbers these days. Travel gives you a new perspective on life. Too many North Americans, I find, are totally parochial in their outlook. They can't accept the fact that there are other people, with different ideas, on this planet of ours. They have no idea of the geography or history or culture of the rest of the world. Nor are they willing to learn. All too often, one sees parties of American tourists in Europe who don't even realise they're in Europe. They travel from one air-conditioned Hilton (or Sheraton or Intercontinental or any other hotel chain that comes to mind; this isn't a sponsored book) to another by air-conditioned coach. They look out of the window at the scenery and discuss the taste of the morning coffee or the rate of exchange on the dollar. They translate everything into American terms, demand some new experience, and complain when it isn't what they're used to back home. They go abroad and are offended when everybody doesn't speak English (even in England, where the English can't speak it properly). They insist on being given hamburgers from Macdonalds for lunch and a steak for dinner. I can never understand why they don't just stay at home and go to a travelogue once a week. They'd have exactly the same benefits at a minute fraction of the cost. It's all a matter of how much psychological comfort they derive from bragging about the miles they've covered, I suppose.

Then there are the good North American tourists, the ones who travel discreetly, who blend into the country they're visiting, who accept a new experience for precisely that, and learn from it about their host countries. These people are prepared to eat new foods, try new languages, walk through new streets, and accept some unfamiliarity. Isn't that precisely the way to develop confidence? When they're presenting themselves with a new challenge almost minute by minute and learning all the time, aren't they doing exactly what I've suggested to become more aware of the world and more interesting people? Sadly, such intrepid travellers are becoming either more self-effacing or rarer. I suppose that, as the boom in conducted bus tours increases, as airline package prices fall, as the status of foreign travel comes within everybody's economic grasp, the temptation to do more for less is strong. It's so easy to go to the travel agent, pay a fixed price, and know that all your needs will be met, in safety and comfort, by the tour guide, the driver, or some other member of the team you've hired for the duration of the trip. Before you take that step, though, ask yourself what it is you're buying. Is it merely a travelogue? Is there a better way to go, a way that will help boost your confidence? Look before you leap!

Well, that's enough about broadening the mind. Let's look now at the related problem of opening it. There's no point in exposing yourself to new ideas if you're not prepared to consider accepting them when they arrive. The secret here is to be ready to consider carefully some new suggestion, from whatever source, before deciding to accept, reject or ignore it. If you constantly take other people's ideas at face value, you'll never develop confidence of your own. If you always reject a new idea without consideration, you'll never be able to grow mentally. Thus, if you want to be comfortable in your mind, consider everything on its merits. If you trust the originator, perhaps you should be more open to acceptance, but even if the suggestion comes from somebody totally unknown or of a type that you've never accepted, that doesn't necessarily mean that the idea itself is bad. Consider it! Again, look before you leap, but don't always refuse to leap.

The end result of this policy will make you a better communicator, a more open person, and hence somebody who'll be sought out by others. The effect will be to boost your own confidence in yourself and, as a consequence, you'll be more comfortable. The entire process of communication, which gives to human beings the sense of belonging, is nothing more than a readiness to exchange ideas with others and to give the same attention to what they're saying as you expect from them. Communication leads to breadth of mind and thus to greater comfort once again.

If you develop these skills, it's obvious that you'll enjoy better relationships with others and, once again, improved comfort. You have to be prepared to give as well as take, though. It's no use bemoaning the fact that nobody will talk to you or listen to you if you're not prepared to reciprocate.

Nowhere is this fact more important than in the most important of human relationships, love. The two partners in any love affair (in the widest possible interpretation of the term, including marriage as well as other less traditionally respected forms) must be prepared to give of themselves completely, without expecting anything in return, if the love is to be worth having. If neither partner is prepared to give, then the relationship is doomed from the start. If one gives and one takes all the time, the affair can never be totally satisfactory (and hence comfortable) because one partner will always feel unhappy or cheated, while the other will always be dissatisfied and making more and more demands.

For any relationship to succeed, too, there has to be complete honesty. Even in the so-called "open marriages," where the partners are totally free to behave exactly as they want and to have countless affairs outside the marriage if they so wish, there can be no real happiness (or comfort) if one partner or the other cheats by hiding some aspect of his or her behaviour.

Honesty with yourself is also important in deriving or maintaining mental comfort, irrespective of the relationships with others. If you do something that's so dishonest you're ashamed of it, the sense of guilt that accompanies you until you've made restitution denies you any hope of comfort. It's not so much the effect your dishonesty has on society that matters but rather the effect it has on you.

The act of giving doesn't mean you have to be totally subservient to others. Nobody can be happy (and therefore comfortable) if he spends all his time being a slave to everybody who asks for some service. If you make yourself into a doormat, somebody will walk over you. It's important to be assertive when you feel that you need to refuse something for your own peace of mind. Everybody needs private time, and if you're constantly doing something for others you'll never have the time you should devote to your own needs and hence your own comfort.

If you're being pushed to carry out some task which you're unwilling to do, refuse firmly but politely. If it suits you, give a reason, but don't feel obliged to explain your conduct if you feel it's nobody else's business. Above all, be polite rather than aggressive. If you handle the situation well, you not only avoid giving offence (and hence risk losing a friend) but you also grow a little yourself and thus improve your confidence and (guess what?) your comfort.

The question of privacy brings to mind the matter of how much privacy we need. There are two extremes to be balanced: the situation where we don't have a minute to ourselves and the one where we never see anybody and are bored to distraction. The amount of privacy you need is a totally subjective matter. I can't tell you how much company you want any more than you can tell me my needs. (Nor would we be prepared to listen to each other if we had the temerity to offer such advice.)

Again, it's a matter of making a decision for yourself and working to accomplish what you feel to be the optimum level, balancing your own (possibly selfish) needs for solitude or comfort against those of the friends who have their own needs but who're prepared to help you meet yours. The opposite of too much company can be either a welcome privacy or a crushing boredom, and it's up to you to decide where the distinction must be made. The difference in the effect on your comfort is, of course, enormous.

Then there is the problem of stress. It's totally impossible for anybody to exist without encountering stress in some form. Psychologists tell us that a certain amount is necessary for us to live productive lives, but it's an undeniable fact that too much stress can lead to severe discomfort, of a mental kind initially but of a physical kind (such as the onset of ulcers, for example) if the stress continues. The best advice I can give is to suggest that you plan your living carefully to minimise its onset. Allow enough time for all you want to do, and if you find yourself rushed, then slow down. You know the kind of thing I mean; it's what your parents used to drum into you when you were a child. Eat proper meals, sleep regular hours, assert yourself when necessary, be prepared to talk out problems, and handle your finances responsibly.

This mention of finances leads us to one of the most common factors influencing the level of personal comfort: the matter of money. It's a well-accepted fact that money (or, to be more precise, the lack of it) is one of the commonest causes of human discomfort. If you don't have enough money to live in accordance with your perceived life-style needs, then you're unhappy. Marriage partners quarrel frequently about financial matters, and guidance counsellors regard such quarrels as possibly the single most influential factor in matrimonial breakdowns. A lack of money can poison the human mind so badly that comfort can't be found for worrying about the payment of bills that constantly come due.

A common way to tackle the problem is to try to increase income, rather than reduce expenditure. The second job, the overtime, the reliance on monetary rewards from speculation or gambling, all begin to dominate the way of life that you accept as normal. The result, all too often, is an increase in misery because the time that should be spent in developing human relationships is given instead to the pursuit of increased wealth. The rich miser, the business tycoon totally out of touch with his family, the pale-faced student carrying a heavy employment load as well as trying to study—all these are familiar characters on the stage of life. The end result often brings misery on the scene. A marriage or a mind may break down, or a human being may decide to end his or her life. Even if the outcome is less drastic, there is still a lot of unhappiness, and thus discomfort, produced for somebody.

What, then, is the solution? We've seen that overwork, though it may solve the immediate cause of the financial shortfall, doesn't bring about any long-term increase in total mental comfort. The only way to assure that, as usual, is to tackle the problem at its roots. The best approach is to assess squarely the reason for the financial difficulty. Is the life-style you seek unrealistically high? Are you influenced by too many other people in your thinking? Is your model of behaviour the local millionaire or a Hollywood star? Are you trying to "keep up with the Joneses" when they're out of your financial league? Only you can decide what the true reason for your problem is, but if you face up properly to it and are totally honest with yourself, the cause will be found and the effort of identifying it will almost certainly bring enough clarity into your thinking that the solution will stand out unmistakably.

When that happens, you've got some latitude in how you tackle the problem. You can decide to move to another neighbourhood, where the standards are less exacting. You can work out a budget (either by yourself, in a family unit, or with the aid of a debt counsellor) that enables you to live within your means by making some adjustments to your life-style. You can even carry on in your present manner but trim all your spending slightly so that money goes the little bit further that will make all the difference. It seems too evident to be worth mentioning, but it's so much simpler if you do all this kind of planning before the financial situation forces it on you. The size of the average personal debt in North America, though, indicates that precious little planning goes into the personal budget of a hefty proportion of the population in general!

Now we come to the matter of physiological comfort. How do we ensure that our bodies are at their optimum comfort level, having got our minds to an acceptable state?

The first need, I'd suggest, is to safeguard your health. No matter how you strive to increase your personal comfort, your efforts are doomed to failure if you're ill. Common sense should tell you to take elementary precautions that enable you to survive life's buffeting. A healthy human being is better able to cope with stress than a sick one. A sensible diet, proper sleep, and a reasonable amount of exercise are all essential to maintaining health.

Diet is a matter of following the guidelines already developed and publicised widely. This isn't a book on food, nor do I claim to be an expert on the subject of nutrition. I leave that for others better trained than I am. However, it's common knowledge that the human body needs protein, starch, minerals, vegetables, dairy produce, vitamins, and all the other essential foodstuffs we read about daily. To deny yourself proper nourishment by grabbing a quick snack that fills your stomach without providing a proper source of dietary intake is a sure way of building up a store of future problems. Get advice on what (and how much!) to eat, and follow it, if you want to remain healthy.

Sleep and exercise, in the same way, have to be planned. There is no point in staying up until the small hours night after night when an important piece of work is due, then trying to catch up on the sleep deficit a week later. The harm to health has already been done. It's not possible to carry forward a sleep deficit for a long time without doing some damage, any more than you can carry forward a dietary deficit without harm.

With exercise, the situation is similar. Too many people assume that exercise benefits can be stored up, like a camel stores water, and engage in one bout of strenuous activity a week. In fact, that is about the most stupid approach you can take. The weekly game of tennis, or squash, or whatever your fad happens to be, can be (quite literally!) a killer. Not only do you always get the aches and pains that delude you into the thought that you're doing yourself some good, but you also can do actual harm. The same overload effect that makes your leg ache is affecting the other muscles of your body, and your heart (in case you didn't realise it) is just another muscle. Unfortunately, the heart doesn't have many nerves attached to it, so it can't tell us when it's getting overloaded. It finally lets us know, in no uncertain terms, that it's had enough by a malfunction that can disable us for a long time or even kill us.

The only sensible approach to exercise is to begin cautiously and increase gradually to the point where you're exercising vigorously on a regular basis. Not only does that avoid placing an undue stress on the heart by sudden loading, but it also ensures that the aching muscles stop aching after the first few weeks. Muscles ache (as we've seen) when lactic acid is generated in them, and regular work enables them to cope with the dissipation of this compound more efficiently, thus eliminating it before it builds up to a painful level. So exercise by all means, but do it in a carefully controlled programme and at least three or four times a week for a minimum time of half an hour per session.

If you want to stay healthy, you should take care not to abuse your body in other ways. The cigarette may look very glamorous when somebody else is smoking it, but the things it does to your lungs, your liver, your arteries and your heart aren't all that funny. A quick shot of something from a needle may give you all kinds of wonderful visions, but they end eventually and you still have to face the realities of life (or kill yourself in the effort to escape it). Dietary fads that depend on half a grapefruit a day for ten weeks and guarantee that you'll lose twenty pounds sound very good in magazines, but the doctor or undertaker won't laugh at your anorexic physique. The current fascination with S and M is interesting to hear about but not too pleasant (or at least, I would assume not; I'm not a devotee!) to experience at close quarters.

We can go on for a long time, getting more and more salacious, but I suspect I don't need to hammer the message home any harder. Abuse of your body, whether for weight loss, sport, or just plain fun, can be a dangerous game to play. For continuing comfort, be aware of the need for caution. This applies also to everyday living and the matter of safety.

Such dangers as corrosive chemicals, household poisons, toxic gases, fire, hot ovens, boiling water, sharp edges, knives, scissors, electrical wiring, medical drugs, and all the other dangerous substances that lurk around a home should be treated with extreme caution. Lights that are too bright or too dim, and sounds that are too loud, should be shunned. Regular medical examinations to prevent the development of problems should be encouraged. Precautions should be taken against extremes of cold or heat when you go outdoors and against allowing sweat to cool down too rapidly after exercise. These latter precautions are usually handled by clothing, a topic which I'll be discussing in more detail shortly.

In terms of physical comfort, clothing is also very important, but, once again, I'll postpone a discussion of its role to a more appropriate part of the chapter. Other physical aspects include taking care not to damage the eyes, ears, or other parts of the body by carelessness. Don't, for example, sniff or taste any unknown substance, and take care that food isn't too hot before eating large quantities of it. The same precautions that we considered in connection with physiological comfort are still applicable. Don't risk injury by foolhardy behaviour. Don't risk straining a muscle (including the heart) by pushing it past its physiological limit. It's true that all these are matters of common sense, but the accident statistics demonstrate only too clearly how often such elementary precautions are neglected or ignored.

When you want to enjoy life to the full, be aware of how important quality of life is for you. Make sensible choices of where to live, how to spend your time, where to place your priorities. If you hate driving, there is no point in buying a home that is an hour's drive through heavy traffic from your work place. Conversely, if you can't stand solitude, it's foolish even to consider taking a job in a sleepy village and living around the corner.

If you're seven feet tall, a subcompact car may leave you a little cramped, but if you can't see over the dashboard (or over the next installment of the bank payment!) after buying the latest version of a giant gas-guzzler of a car, maybe you should have gone for something smaller. If you dare not show your face at the office because your latest garment purchase is too loud, or too sedate, then possibly you should consider changing your shopping habits. If the food you eat is tasteless or too expensive, the same remedy is needed.

Always remember, though, that what you do in making your choice may have some effect on the rest of the population. You should be thoughtful for other people's needs, especially those near to you. It's hardly fair to insist on living in the middle of a big city for convenience to your job if your spouse is terrified of noise, or traffic, or any of the other things that go along with city living. Nor is it fair to insist on eating out three times a week if somebody else (like your spouse) responsible for the budget has problems in making it stretch far enough.

In the matter of consideration for others, remember too that they have to share your environment, whether on a small or a large scale. My personal dislikes (as you may have gathered from comments I've already made!) include smokers who insist on puffing their foul effluent into my air, and those kind friends of humanity who leave their refuse behind to let me know what they had for their picnic lunch in the beauty spot I'm trying to enjoy. Every small contribution you make, whether to pollution or to its control, can have a double cumulative effect. In the first case, anybody following your example will multiply what you've done, for good or bad, and in the second case you'll become so immune to the sight of scattered rubbish that the appreciation of beauty will be dimmed without your awareness of the fact. So resist the temptation to pollute the planet, in whatever form your crime may take, and enjoy what is available in nature. Somebody else may even do the same kindness for you!

Now we come to the tricky subject of fashion. I've already expressed my opinions on this subject in an earlier chapter, but I recognise that, for some people, my ideas would be totally unacceptable because their comfort level couldn't tolerate the resulting loss of prestige. It's only fair, then, to look at what can be done to help in such cases. I think it's reasonable to ask you, before you rush out to buy a new wardrobe, to consider the cost of your actions, not only in financial terms but also in the other areas we've been discussing. The world is slowly realising the enormity of this type of cost, and such steps as banning the export of garments made from the skins of endangered species are being taken. Consider the social cost of your purchase, then, and only buy if you can honestly claim that your need is great enough.

Now, what other factors should you take into account? Money clearly has to be considered by those of us not fortunate enough to have no need to worry about the cost. The solution recommended by all the fashion magazines is to buy wisely. Choose coordinated colour schemes and match garments so that you can vary the outfit by selection from a fairly small range of units. For the men, this can mean buying a suit with an eye to using the jacket as a sports coat with other trousers. For the women, the same sweater can be used with a skirt or with slacks on different occasions. Check your wardrobe and see how many different combinations of this kind you can find. If there are very few, that is an indication of lack of planning before you shop.

When you set off to buy clothes, then, there are several precautions you can take before leaving home. Decide first what aspect of your personality you wish to express. It's no use coming home with a wildly gaudy outfit if you're looking for a way to impress the local banker into lending you money. On the other hand, a severely tailored skirt and blouse is hardly the thing to attract a man you're trying to seduce into a wicked weekend with a bunch of other swingers.

Next, decide on the colour, style, fit, texture or other characteristic you're seeking. It's no use rushing into the nearest clothing store and grabbing half a dozen articles at random because they match the rainbow you just saw or because the nubby texture reminds you of the dog's coat when it needs brushing. Have a clear idea of the type of garment you're after and of what you'll need to wear with it. If the shade is critical, take one of the garments it has to complement along to the shop with you, rather than trying to remember a subtle colour you have hanging in the closet at home. Remember, too, that colours can differ considerably in different lights. Examine the two garments together in daylight as well as in the shop, and decide whether you can tolerate the difference for the end use intended. If you're going to wear the outfit only for evenings, it may not be critical that the colours don't look as good together in daylight.

For the last word on this topic of fashion, let me make a special plea for a hopeless cause. If you really want to be comfortable, then choose physical comfort in preference to psychological comfort. I know I'm probably beating my head against a brick wall, but I really do assure you that, in the long run, you'll be glad you took my advice. The anguish of wearing a garment which is physically uncomfortable, merely because it happens to be in fashion at the moment, can bring misery when your feet, or your varicose veins, or your twisted ankle, or any of the other symptoms of discomfort that you'll eventually have to tolerate, start to afflict you. I realise that I'm wasting my time in pleading with you for this small concession, but at least the paragraph swells the chapter and thus nets both the publisher and me financial benefits!

Now we turn to the home. The same kind of criteria are also important here, in general. The budget considerations have to be taken into account and so do the personal tastes of the people living there. You've got to balance your income, perceived life-style needs, taste, space requirements and living costs to decide how small or large the home should be, how much you can afford to spend on furnishing it, where it should be located, and how long you'll need to pay for your preferences. Don't forget also, of course, that a house needs constant expenditures for maintenance, taxes, insurance, repairs, redecorating, and all the many other pleasures of the homeowner.

What, then, is a logical order to adopt in making decisions of this kind? The structure of the building itself should, I think, be your first concern. Is it sturdy, or likely to collapse in anything stronger than a gentle breeze? Is it well insulated, or can you see daylight through the cracks in the walls? Is it big enough for your family, or will you have to eat dinner, sit to watch television, or use the bathroom in relays? Is it in a desirable location, or will you have to hire security guards to come around once a week to protect you when you put the garbage out? Will the children be happy there, or is the street crawling with child molesters, drug pushers, and similar pleasant and upright citizens of good repute?

On the naive assumption that all these objections can be met and you find yourself in a home, what needs to be done to make it comfortable? Physically, it needs to be maintained at a pleasant temperature, which may require heating or air-conditioning arrangements. For fuel, the choice lies between electricity, gas, wood, oil or coal and is influenced by local conditions. Check carefully to decide which of them is best for you, using all the information resources (such as libraries, municipal authorities, or long-term residents) that you can find. Make sure that the insulation is sound, particularly if you live in a part of the world that is as cold as my neck of the woods. For those living in a city or other noisy environment, the provision of double-glazing for all windows will be of benefit in reducing noise as well as in preserving heat. For comfort of a different kind, insect screens may well be a necessity for comfortable summer living without the risk of being bitten incessantly.

Then we come at last to the matter of the aesthetic aspects of comfort. Personal tastes are so important here that it's impossible to lay down rules. All that can be done is to suggest that, if you feel at all dubious about your ability to create a harmonious decor, you should ask for advice. The advice may be as simple as getting the opinions of a few other people before buying furniture, wallpaper, etc., or (if the budget permits) it may involve hiring a professional interior decorator to advise you on your choice of colours, styles, and so on.

Once your colour scheme and harmony are decided, you can begin the task of transforming the house into a home. Plan carefully, following the scheme you've established. Be realistic in your aims; there is little point in being so ambitious that you'll have to wait fifty years before your dream is finally realised. Even if you survive to see the dream accomplished, the first stages will be so dated that you'll have to start all over again before you can possibly live in the hideous place! Take a room at a time, plan it carefully, work at a pace to suit your energy and budget, and divide the process into small (and therefore realistic) steps that can be accomplished steadily. In that way, you'll get a continual sense of achievement as you finish each step and the end result won't seem too far away.

In considering work conditions, you should make your choice (assuming you have any in these times of high unemployment!) with certain factors in mind. Balance the need to earn money with the job satisfaction you expect to get. If you don't care about work but only want to get as much money as possible for the least effort and are prepared to tolerate virtually any condition to this end, there is not too much more to be said. The activity of trade unions makes it very unlikely, at least in the developed countries of the West, that you'll be badly exploited.

If, on the other hand, you need fulfilment from your job, then you may have to sacrifice financial rewards. It's a sad but true fact that the really interesting jobs are usually paid at the lowest rates when compared with others needing the same level of training or experience. You've only to think of the poor struggling university professors, with perhaps the most challenging and demanding career in the world and with a miserable pittance for their reward, to realise the fundamental truth of this statement. The fact that I happen to be a professor, of course, is totally unrelated to the validity of the argument.

In your work, you can reasonably expect to have enough privacy or company to allow you to work at an efficient level. What you can't expect is complete leisure to sit and chat in a friendly atmosphere of bonhomie all day without getting any work done (unless you're a politician, or an administrator of some other kind, who, as we all know, is required to do this kind of thing by the conditions of employment). You can also expect your employer to take all reasonable precautions to ensure that your working environment is a safe one, both physically and mentally. If it's not, the mechanism for complaint is usually in place and you should find or use it.

You may have to accept a compromise in such matters as temperature, humidity, lighting or sound levels, and other environmental factors, either because the arrangements necessitate a democratic compromise or because the nature of the employment dictates an economic one. In either case, there are bounds which are acceptable and, if conditions stray outside them frequently or for too long, you've every right, once again, to complain and seek restitution. Naturally, some aspects of the work may also require you to observe regulations designed to preserve safety, either for you or for somebody else, and you must expect to follow reasonable guidelines of this kind.

In transportation comfort, you're again faced with a number of choices. Cost, as usual, will be a major determining factor, one which you must balance against convenience. It's very nice to have a private vehicle to take you everywhere, but you've got to realise that you'll pay heavily for the luxury. If you live in a large city, where public transportation is convenient and where heavy traffic or difficult parking makes driving an automobile a nightmare, perhaps you shouldn't even consider buying a car. On the other hand, you may live in the country or have to move around frequently as part of your work, in which case the experience of waiting for buses or trains may drive you insane. The mental cost, as well as the financial one, has to be considered when making your decision.

What you should be careful to do is distinguish need from desire. A car that enables you to get to work in half an hour instead of two hours is a necessity, but one that enables you to become the Don Juan of the county, no matter how much satisfaction (and comfort of this or other nefarious kinds!) you get from the result, must be considered a luxury. In calculating the financial cost of your conquests, you've got to include repair, maintenance, fuel, oil, tax, licence and insurance (and, if you really are a Don Juan, no doubt you'll be busily collecting sundry fines to boot). The total cost of running a car (in addition to the cost of financing it if you didn't pay cash) can amount to several thousand dollars a year, a fact which escapes the attention of most would-be motorists until they sit down two or three years after the glow of first ownership has evaporated and figure out the arithmetic. The mental discomfort of a financial burden may be a high price to pay for the mental (and physical) comfort of the vehicle.

Let's assume that your car purchase is genuinely a necessity after all the evidence has been carefully weighed. You now have to make the decision about what type of car to buy. Do you get a large or a small one, and buy it new or used? In general, the advice I would give is to get the best car you can afford. If you can afford a new one (after taking running costs into account), then get it. The peace of mind will be more than worth the expense as long as you don't suffer financial hardship in consequence. In the matter of size, I'd suggest that your choice will be governed by your own feelings of personal insecurity. The possession of a huge vehicle gives the owner a sense of importance (and hence comfort) out of all proportion to its true worth. If your self-esteem needs that kind of boost, then spend the money and bask in the envy of others.

If you're already convinced of your own self-worth, however, I'd recommend a smaller automobile, because the comfort of lower running costs will more than repay your trivial loss of prestige in the long run. In my time as a resident of North America, I've tried both large and small cars, and gained the comfort rightfully bestowed by each one. I'm currently driving a small one and enjoying immensely the financial advantages, but my last car was a large one and was much more comfortable to drive and own. One lesson I've learned, though, is to buy a vehicle that is physically comfortable. It's pointless to take the economy comfort to extremes and buy a miniature car if you're over six feet tall as I am. The resulting physical discomfort will never be absent and will wipe out all the advantages of psychological comfort gained by your sacrifice. Luckily, my lesson was driven home (if you'll excuse the expression) when I was a student, in a car that cost me the Sterling equivalent of fifteen dollars, so the mistake wasn't too expensive.

Now, what about the reader who has decided, for whatever reason, to forego the luxury of private ownership? What are the comfort factors that must be pursued in using public transport?

Your most important decision is that of choosing the mode of travel. Distance will usually determine this to some extent, together with cost. If your destination is overseas, you're almost (but not quite) restricted to air travel. Ships do still sail, and if you can afford both the time and the fare, I'd go that way every time if I were you. The relaxation is more than worth the cost. If time or money is limited or if your journey is overland, then you've little option but to fly. The alternative, a trip by bus or rail, is realistic for journeys of up to five hundred miles or so, but beyond that the mental strain of the prolonged journey and the cost of meals or rooms en route are both too high to compensate for the difference in fare.

For journeys of moderate length, I personally prefer trains, even at the higher cost, mainly because I tend to get travel sick in a bus where the smoking section isn't segregated. I can also read on a train but find it impossible on a bus. I even choose to travel overnight on a train (with sleeping accommodation) in preference to flying when I visit Montreal or Ottawa, a distance of about five hundred miles, mainly on comfort grounds. The train cuts out the need to get up at the crack of dawn to rush to the airport and avoids, too, the need to take airport transportation at each end of the journey. As far as I can tell, I'm almost alone among my business associates in this preference, but I recommend it if you want a relaxed trip!

For short journeys, the choice lies between a bus or a taxi (unless the trip is so short that you can walk or cycle). Cost and urgency both govern the decision, though there is little to choose in time economy in a big city (where there may well be an underground subway, Tube, Métro, or its equivalent that is better than either). I tend not to use taxis, mainly because I always try to leave myself plenty of time to get somewhere and because I prefer to walk if at all possible. My wife tells me that I'm also too cheap, and it's rude to argue with a lady.

Having selected the preferred mode of transport, you should next consider the choice of seat. On an aircraft, I always travel at the cheapest available rate, on the grounds (whatever my wife may say) that flying is uncomfortable no matter where you sit and that I can make better use of the money saved in finding relaxation (and better comfort) when I get off the plane. I ask for a non-smoking seat (and get very unpleasant if denied it!) by the window, on the side away from the sun to avoid the discomfort of being overheated and dazzled. This entails an elementary knowledge of geography and a more intimate knowledge of my destination, so that I know where the sun will be during the entire flight, but the effort of doing that bit of homework is more than repaid by the increased comfort.

My choice of a window seat is also governed by comfort of another kind. In it, I'm not constantly being asked to stand so that the dear old lady with the nervous system or weak bladder can pass me on the way to pass something else. Nor am I at risk of being jostled by children or stewardesses with precariously balanced pots of tea and coffee as they trundle up and down the aisle. I've had more fluids, and of a wider variety, spilled on me by more passengers and employees than I care to recall during my years of flying, and my fussiness about seating is the end result of many lessons. The window seat also reduces the risk of boredom (since there is always something to look at, even if it's only the changing shape of clouds, apart from the passengers) and eliminates the need to talk to the garrulous fellow-traveler if you don't feel inclined to converse.

There are, it's only fair to mention, people who prefer not to take a window seat for the simple reason that they're nervous about flying. As long as they can't see outside, they don't feel the fear (and hence one of the most disturbing discomfort aspects) of flying. Those of us who're not troubled by this phobia find it hard to understand, but I've known people who suffer absolute agonies of terror from the instant they enter an aircraft and don't shed their fear until they reach the terminal building after the flight has ended.

In other modes of transportation, the choice of seat is less critical. In a train, apart from the difference between classes and the window vs. aisle decision, there is little to choose between seats. They tend to be more spacious, and it's not usually difficult to see out of the larger window even if you're in the seat away from the view. Smoking compartments, as already discussed, are also usually segregated.

On a bus, when I'm forced to use one, I try to avoid the seats over the wheel and at the back, because the vibration there is greater. Apart from that (and the perennial smoke problem, of course) there is no real difference in bus seating as far as I'm concerned because the scenery is so blurred by vibration that it's not worth looking out of a window. It may have escaped your notice, but I have a slight bias against bus travel.

The last aspect of transportation comfort I want to examine is that of time. I always try to arrive at my place of embarkation in ample time to board, so that I'm not worried about missing my flight, train, boat, etc., and nearly drive my family or friends (who have this peculiar and vindictive wish to break all speed limits getting to the embarkation point, where they climb aboard the vehicle just as the thing is starting to move) insane in the process. There ought to be a happy medium, but I haven't quite managed to find it yet. At least I've never yet missed a vital connection, though my luggage (as already mentioned) missed a plane on one occasion and that certainly caused me discomfort!

Too much time spent on a journey can cause fatigue and boredom, particularly if you have to travel with young children. It's well worth the effort of carrying a few games to distract them from the tedium if your children aren't old enough to amuse themselves. When our own children were very young, we used to carry a selection of small puzzles or games with us whenever we flew across the Atlantic. At the end of the trip, the children were always still reasonably well behaved (unlike some poor children we once saw who screamed with boredom throughout the entire journey as a result of their parents' lack of forethought), even though I was invariably a physical wreck as a result of all the effort expended in keeping them amused.

To make one last point regarding travelling with children, it's also not a bad idea to carry a small first-aid kit with you. The stewardess on a plane always has one, but there are other vehicles and it's my experience that a child will always cut its finger when the nearest public first-aid kit is empty or has been stolen.

Finally, I come to the most difficult part of the entire book: the matter of expressing personal opinions about the many textile products I've mentioned in the foregoing pages. Perhaps it's because my training as a textile scientist makes me much more aware of all the possible reasons for using different types of textiles, but I always feel reluctant to recommend one particular type over another. I know my own preferences and I know why I have them, but I hesitate to impose them on other people with all the authority of the printed word. However, I've never (well, almost never!) been one to shirk a duty, so I'll merely repeat my earlier word of warning that the opinions I express are personal ones and press on with the task.

With that proviso in mind, it's fair to say that some of my opinions are based on solid scientific facts that can't be denied. The natural fibres, for example, will always be more comfortable (in the matter of moisture and thermal transmission) than the currently available synthetic ones because of the differences in molecular structure in the two cases. Conversely, the same synthetics will be easier to maintain, for reasons relating to the same molecular structures, than the natural fibres, and easy-care characteristics may well be of greater importance in the comfort needs of some readers.

For my preference, I would choose 100 percent cotton for underwear and shirts, in all seasons. In summer, I try always to wear cotton or cotton/polyester (with not more than 50% of the polyester) outer garments for reasonable appearance with optimum physiological comfort, and cotton or lightweight wool socks.

In winter, I wear 100 percent wool suits, trousers, and sweaters. For outer wear, in dry weather I wear a woollen overcoat and, in rain or snow, a polyester/cotton or nylon/cotton wind- and water-resistant topcoat. In extremely cold conditions, I'm not ashamed to be seen in a peculiar woollen garment which combines the attributes of a balaclava helmet with those of a scarf, covered with a water-resistant hood for additional protection. On my feet, I wear heavy woollen socks and climbing boots, ski boots, or knee-length boots, depending on the kind of activity I plan, and leather mittens with a fur lining. The equivalent garments for the ladies are fairly easy to identify.

The principles on which a good winter clothing assembly is designed aren't difficult to recognise. The underwear needs to absorb perspiration so that sweat doesn't condense near the skin. The system should consist of many layers of fabric (shirt, jacket, linings, interlinings, insulating layers and outer layer, for example) and should be covered at the outer surface with a wind-resistant fabric (which may also with advantage be water resistant). Warmth, as we've already seen, depends on air entrapment, and the natural fibres (with their convoluted surfaces) are better at this feature than the cylindrical, synthetic ones. In addition, of course, the ability of the natural fibres to absorb more moisture makes them better able to deal with perspiration.

The textiles used in the home don't lend themselves quite as readily to such clear distinctions. In the first case, the matter of absorbency and heat retention are usually less critical, while the appearance-retention properties are of more importance. For a carpet (probably the most expensive home furnishing), the fibre type, yarn twist, pile height, pile density, backing material and underpadding chosen are critical. If the fibre absorbs dirt readily, the carpet will soon look dingy. If the yarns are too tightly twisted, the carpet surface will be harsh, while too loose a twist causes rapid flattening which looks unsightly. A pile which is too short will wear away quickly, while one that is too long will, again, flatten; so will a carpet in which the pile density (the number of pile yarns per unit area) is too low. Too high a pile density, on the other hand, as well as increasing the cost immensely, can lead to accelerated wear because the yarns can't "give."

My own favourite carpet of modern type is a hard-twist acrylic which has given us good service for years. This particular style seems not to be obtainable any more, but a hard-twist polyester should be an acceptable substitute. I prefer to avoid nylon, because it develops a shine quickly (and hence looks worn almost from new, though it will probably last for many years afterwards), but the new types of nylon are supposed to avoid this problem. I'll test them one of these days, but the results won't be available, I'm afraid, for inclusion in this book unless it goes into three or four reprints or new editions. Wool carpeting, once the mainstay of the industry, is rapidly disappearing on account of high cost, as also are Axminster and Wilton woven ones, now universally replaced by tufted carpeting. We have an Axminster, woven from wool, which has served us well for twenty-five years and still looks almost new. I wish I could make the same claim for most of our tufted ones!

The optimum pile height, I'd suggest, is about five millimetres. If you want a luxurious feel, your best plan is to buy a thicker underpadding, which will cost you less and give you far better service than a carpet of higher pile. To determine whether the carpet has an acceptable pile density, bend the face of the carpet backwards until you can see the ground on which the yarns are tufted. The jute (or other material) should be clearly visible, but there should be no wide spaces between adjacent tufts. If the carpet is intended for use where water or other liquids are spilled frequently (and particularly for outdoors), then insist on 100 percent polypropylene, which is rot and stain resistant to a far greater extent than all other fibre types. In the case of outdoor carpeting, the underpadding must also be of polypropylene, or odour problems will arise.

Upholstery needs to resist staining and wear. It should also not feel clammy if you sit on it in hot weather. I prefer a linen treated with a stain-repellent finish, but a polyester fabric which isn't too tightly woven is a good choice also. Nylon, in my experience, again becomes shiny (and hence unsightly) very quickly, so I'd avoid it. I'd also tend to avoid cotton or rayon, since they lose their appearance qualities faster than linen. Wool is an excellent upholstery fabric in most cases, though stains tend to be more difficult to remove because the fibre is damaged by bleaching agents. The natural fibres, of course, are absorbent when sat on, which makes them preferable for summer comfort.

Curtains pose a different problem. They should be strong and keep their new appearance but not be harmed by sunlight. They should also be fire resistant, since household fires often spread from one room to another via the vertical curtains. Glass fibres were popular (for the latter reason) several years ago, but they tend to be rather brittle and break easily, looking unsightly. Polyester treated with a flame-resistant finish or modacrylic (which is inherently flame resistant) are preferred nowadays. I'd avoid cotton, rayon or silk, which are degraded by sunlight, and wool, which is difficult to wash (as also is silk) and tends to lose colour fairly easily on prolonged exposure to light.

For bed linens, a cotton/polyester blend is ideal. It's strong enough to resist the penetration of a long toenail but comfortable enough to eliminate sweating in all but the hottest of climates. Blankets used to be all wool, but laundering difficulties coupled with the advent of central heating have brought about a change to lighterweight cellular goods, usually of acrylic, which are perfectly satisfactory. For an eiderdown or duvet, a polyester cover is satisfactory if you remember to remove the furnishing before going to sleep, to avoid perspiration condensation. The newer feather-filled articles are more satisfactory in this respect because they allow moisture to be transmitted freely, so reducing or eliminating this disadvantage.

The choice of fibre content for table linens depends greatly on the use intended. For a formal dinner, there is still no substitute for top-quality linen, though a polyester/cotton cloth gives almost as good an effect at a much lower cost. Where children tend to eat frequently, a plastic cloth (transparent if you wish to show the real cloth beneath) is a wise investment even though it looks a bit ugly.

Linen is also best for the towels that dry the dishes, since it absorbs water extremely well, washes easily, and lasts for an adequate time. For towels used to absorb water from the body, though, choose cotton. It can be made with a much softer hand than linen, is absorbent, and retains its strength well when wet. For best absorbency, incidentally, buy a pale colour rather than a dark one, since the large quantity of dyestuff present in a dark shade will have occupied some of the sites at which water could otherwise have been absorbed.

Well, there you have my preferences. They give me the optimum comfort possible under any normal circumstances, and I hope the suggestions I've made, not only in this chapter but also in the rest of the book, will enable you to make reasoned decisions that help you to be as comfortable as possible, given the circumstances under which we live this life of ours. There is no point in being needlessly uncomfortable; we only have one life to live and we may as well make it as pleasant as we can for as long as we can. Don't forget that your body and mind are constantly striving for comfort all the moments of your life. It's up to you to do all you can to help them. No point in fighting it; that will just make you uncomfortable!

And there, I'm afraid, we come to the end of our journey together through the intricacies of human comfort. I hope that I've given you some indication of the complexities of the subject and perhaps opened your eyes to its importance. We've only managed to skim the surface of some aspects of the topic because of its enormity, but if my suggestions make you think a little more deeply than you would otherwise have done, and if they enable you to recognise (and correct) any comfort problems which you may encounter, then the task of writing has been a rewarding as well as a pleasant one. Who knows, if you recommend the book to enough people, it may also have been a profitable one? Needless to say, this is a possibility which makes me feel very satisfied (and hence very comfortable) indeed! Thank you for staying with me to the bitter (or sweet, depending on how the sales pattern actually goes) end!

BIBLIOGRAPHY FOR FURTHER READING

THERE ARE many books which will be of use in further study of the subject of human comfort, and I give in the following list only a few of them. These are selected as examples of sources written in language simple enough to be understood by the educated lay reader, yet not so trivial that they produce exasperation. Any reputable library will be able to find a dozen better books for you on any of the subjects covered by them, so go ahead and do your own literature search if you prefer. The more lazy readers can, however, ask for the following titles in the knowledge that they are reasonable ones:

Aiken, L.: *Later Life*. Philadelphia, Saunders, 1978.

Alexander, P.R.: *Textile Products*. Boston: Houghton Mifflin, 1977.

de Armond, S.J., Fusco, M.F., and Dewey, M.M.: *Structure of the Human Brain*: A Photographic Atlas. New York, Oxford, 1976.

Atrens, D.M., and Curthoys, I.S.: *The Neurosciences and Behaviour*. New York, Academic Press, 1982.

Barlow, H.P., and Mollen, J.D., (Eds.): *The Senses*. New York, Cambridge, 1982.

Barr, M.L., *The Human Nervous System*. New York, Harper, 1979.

Bell, G.H., Emslie-Smith, D., and Paterson, C.R.: *Textbook of Physiology*. Churchill-Livingstone, Edinburgh, 1980.

Beranek, L.L.: *Noise Reduction*. New York, Krieger, 1980.

Birren, J.E.: *Relations of Development and Aging*. Springfield, IL, Charles C Thomas, 1964.

Bouhuys, A.: *Breathing*. New York, Grune and Stratton, 1974.

Brown, E.L., and Deffenbacher, K.: *Perception and the Senses*. New York, Oxford, 1981.

Brown, T.S., and Wallace, P.M.: *Physiological Psychology*. New York Academic Press, 1980.

Campbell, H.J.: *The Pleasure Areas.* New York, Delacorte, 1973.
Carlson, N.R.: *Physiology of Behaviour.* Boston, Allyn & Bacon, Inc., 1977.
Carlsoo, S.: *How Man Moves.* New York, Heinemann, 1975.
Cherniack, R.M., Cherniack, L., and Naimark, A.: *Respiration in Health and Disease.* Philadelphia, Saunders, 1972.
Chrenko, F.A.: *Bedford's Basic Principles of Ventilation and Heating.* London, Lewis, 1974.
Comroe, J.H.: *Physiology of Respiration.* Chicago, Year Book Medical Publishers, Inc., 1965.
Davis, H., and Silverman, S.R.: *Hearing and Deafness.* New York, Holt, Rinehart and Winston, 1970.
Corbman, B.P.: *Textiles: Fiber to Fabric.* New York, McGraw-Hill, 1975.
Dreyfuss, H.: *Designing for People.* New York, Viking, 1974.
DuBois, E.R.: *Fever and Regulation of Body Temperature,* Springfield, IL, Charles C Thomas, 1948.
Duncan, T.: *Physics.* London, John Murray, 1982.
Durrant, J.D., and Lovrinic, J.H.: *Bases of Hearing Science.* Baltimore, Williams and Wilkins, 1977.
Eyzaguirre, C.: *Physiology of the Nervous System.* Chicago, Year Book Medical Publications, Inc., 1969.
Fanger, P.O.: *Thermal Comfort.* New York, McGraw-Hill, 1973.
Farris, M.T., and Harding, F.E.: *Passenger Transportation.* Englewood Cliffs, NJ, Prentice-Hall, 1976.
Folklow, B., and Neil, E.: *Circulation.* New York, Oxford, 1971.
Fourt, L., and Hollies, N.R.S.: *Clothing: Comfort and Function.* New York, Dekker, 1970.
Gale, E.: *From Fibres To Fabrics.* Mills and Boone, 1978.
Geldard, F.A., *The Human Senses,* 2d ed. New York, Wiley, 1972.
Gelfand, S.A.: *Hearing.* New York, Dekker, 1981.
Gerber, S.E.: *Introductory Hearing Science.* Philadelphia, Saunders, 1974.
Gloag, J.: *A Social History Of Furniture Design.* Cassell, 1966.
Gordon, G. (Ed.): *Active Touch.* Pergamon, Oxford, 1978.
Grandjean, E.: *Ergonomics of the Home.* London, Taylor and Francis, 1973.
Hardy, J.D., et al. (Eds.): *Physiological and Behavioural Temperature Regulation.* Springfield, IL, Charles C Thomas, 1970.
Hoffman, A.M.: *The Daily Needs and Interests of Older People.* Springfield, IL, Charles C Thomas, 1970.
Hollen, N., Saddler, J., and Langford, A.: *Textiles.* New York, Macmillan, 1979.
Jones, W.P.: *Air Conditioning Engineering.* London, Arnold, 1973.
Joseph, M.L.: *Essentials Of Textiles.* New York, Holt, Rinehart and Winston, 1980.
Joseph, M.L.: *Introductory Textile Science.* New York, Holt, Rinehart and Winston, 1977.
Kao, F.K.: *Introduction to Respiratory Physiology.* New York, Elsevier, 1974.
Kenshalo, D.R. (Ed.): *The Skin Senses.* Springfield, IL: Charles C Thomas, 1968.
Kerslake, D.M., *The Stress of Hot Environments.* New York, Cambridge, 1972.

Kreider, J.F., and Kreith, F.: *Solar Heating and Cooling*. New York, McGraw-Hill, 1977.

Kryter, K.D.: *The Effects of Noise on Man*. New York, Academic, 1970.

Lamb, J.F., Ingram, C.G., Johnston, I.A., and Pitman, R.M.: *Essentials of Physiology*. Boston, Blackwell, 1980.

LeBlanc, J.: *Man in the Cold*. Springfield, IL, Charles C Thomas, 1975.

Lindsay, R., et al.: *Body Mechanics*. Dubuque, IA, W.C. Brown, 1979.

Lippold, O.: *Human Respiration*. San Francisco, W.H. Freeman, 1968.

Livingston, W.K.: *Pain Mechanisms*. New York, Plenum, 1976.

Lyle, D.S.: *Modern Textiles*, New York, Wiley, 1982.

Lyle, D.S.: *Performance of Textiles*. New York, Wiley, 1977.

Mayer, J.: *Human Nutrition*, Springfield, IL, Charles C Thomas, 1972.

McGraw-Hill Encyclopaedia of Science and Technology, New York, 1982.

Melzack, R.: *The Puzzle of Pain*. New York, Basic, 1973.

Melzack, R., and Wall, P.: *The Challenge of Pain*. New York, Penguin, 1982.

Moncrieff, R.W.: *Odours*, New York, Heinemann, 1970.

Moriarty, D.M.: *The Loss of Loved Ones*, Springfield, IL, Charles C Thomas, 1971.

Murrell, H.: *Men and Machines*, New York, Methuen, 1976.

Myklebust, H.R.: *The Psychology of Deafness*. New York, Grune and Stratton, 1960.

Paul, J.K.: *Solar Heating and Cooling*. Park Ridge, NJ, Noyes, 1977.

Poulton, E.C.: *Environment and Human Efficiency*. Springfield, IL, Charles C Thomas, 1970.

de Reuck, A.V.S., and Knight, J. (Eds.): *Touch, Heat and Pain*, Churchill, London, 1966.

Rosenberg, M.E.: *Sound and Hearing*. London, Edward Arnold, 1982.

Ross, G.: *Essentials of Human Physiology*. Chicago, Year Book Medical Publications, Inc., 1978.

Scharf, B. (Ed.): *Experimental Sensory Psychology*. Glenview, IL, Scott, Foresman, 1975.

Schmidt, R.F. (Ed.): *Fundamentals of Sensory Physiology*. Vienna, Springer-Verlag, 1981.

Schubert, E.D.: *Hearing: Its Function and Dysfunction*. New York, Springer-Verlag, 1980.

Schwartz, M.: *Physiological Psychology*, Englewood Cliffs, NJ, Prentice-Hall, 1978.

Seller, W.A.: *Body Temperature: Its Changes with Environment, Disease and Therapy*. Springfield, IL, Charles C Thomas, 1952.

Shephard, R.J.: *Men at Work*. Springfield, IL, Charles C Thomas, 1974.

Sinclair, D.: *Mechanisms of Cutaneous Sensation*. New York, Oxford, 1981.

Slater, K.: *Comfort Properties of Textiles*. Manchester, The Textile Institute, 1977.

Sloan, A.W.: *Man in Extreme Environments*. Springfield, IL, Charles C Thomas, 1979.

Smith, B.F., and Block, I.: *Textiles in Perspective*. Englewood Cliffs, N.J., Prentice-Hall, 1982.

Stout, E.E.: *Introduction To Textiles*. New York, Wiley, 1970.

Taylor, R.: *Noise*. New York, Pelican, 1970.

Thompson, R.F.: *Introduction to Physiological Psychology*. New York, Harper & Row, 1975.

van Toller, C.: *The Nervous Body*. New York, Wiley, 1979.

Tortora, P.G.: *Understanding Textiles*. New York, Macmillan, 1982.

Weber, E.H.: *The Sense of Touch*. New York, Academic, 1978.

Wingate, I.: *Textile Fabrics and Their Selection*. Englewood Cliffs, NJ, Prentice-Hall, 1970.

Wooldridge, D.E.: *The Machinery of the Brain*. New York, McGraw-Hill, 1963.

Yost, W.A., and Nielsen, D.W.: *Fundamentals of Hearing*. New York, Holt, Rinehart and Winston, 1977.

INDEX

B

O

P